Current Trends in Geomathematics

COMPUTER APPLICATIONS IN THE EARTH SCIENCES
A series edited by Daniel F. Merriam

Current Trends in Geomathematics

Edited by
Daniel F. Merriam

Endowment Association Professor of the Natural Sciences
Department of Geology
Wichita State University
Wichita, Kansas

PLENUM PRESS • NEW YORK AND LONDON

Library of Congress Cataloging in Publication Data

Current trends in geomathematics / edited by Daniel F. Merriam.
      p.   cm.
      "Proceedings of papers presented at sessions sponsored by the International Association for Mathematical Geology at the 27th International Geological Congress in Moscow, USSR, August 1984"—T.p. verso.
      Includes bibliographies and index.
      ISBN 0-306-43087-8
      1. Geology—Mathematics—Congresses. 2. Geology—Data processing—Congresses. I. Merriam, Daniel Francis. II. International Association for Mathematical Geology. III. International Geological Conference (27th: 1984: Moscow, R.S.F.S.R.)
QE33.2.M3C87   1988   88-29383
550′.1′51—dc19   CIP

Proceedings of papers presented at sessions sponsored by the
international Association for Mathematical Geology at the
27th International Geological Congress in Moscow, USSR, August 1984

Dedicated to

Geoffrey W(illiam) Hill
1928–1982
IAMG vice president, 1976–1980
ami, collègue, et monsieur

PREFACE

Since founding at the 23rd International Geological Congress in Prague in 1968, the International Association for Mathematical Geology has organized sessions in conjunction with the Congress. The 27th IGC in Moscow was no exception and the IAMG again held sessions and assisted the Congress in organizing Section 20 - Mathematical Geology and Geological Information (D. F. Merriam, D. A. Rodionov, and R. Sinding-Larsen, conveners). All together 128 abstracts were published in the technical proceedings.

Several of the papers were published prior to the Congress, others were not available, and others deemed not appropriate for publication in this volume. This collection then contains those papers available and representative of the sessions. The collection is truly international with contributions from Canada, China, France, Poland, the UK, USA, and USSR. They are representative of the state-of-the-art as of the early 1980s in a variety of fields.

The application of geomathematics/geostatistics to geological problems has been hastened by the availability of computers. These papers reflect that orientation - most of the results would not have been possible without the use of computers. Most of the approaches utilize techniques readily available and adapted to solving geological problems - simulation, image analysis, decision theory, fuzzy sets, etc. However, one area, that of geostatistiques which includes Kriging, has been designed especially for use by earth scientists of the French school to solve geological problems.

This volume is the fifth in a series on "Computer Applications in the Earth Sciences" that was initiated in the late 1960s. The series was started as an outlet of proceedings of meetings which contained papers of interest to those in the field of

geomathematics/geostatistics/computer applications. Where
possible and appropriate, the proceedings of meetings with this
emphasis are published - including this volume of the IAMG
sessions at Moscow.

I want to thank the authors of manuscripts included in this
volume - they helped ease the preparation of the material with
prompt replies to questions and problems of production.
Several anonymous reviewers assisted in processing the papers.
Mrs. Linda Spurrier of the Department of Geology at WSU helped
with the typing, proofreading, and indexing; the final copy was
typeset by Mrs. Terry Hanley of Falls Church, Virginia; and J.
Thomas Hanley kindly read the final copy. Ms. Patricia M. Vann
of Plenum Publishing Corp. made arrangements for publication.

So here are the results of our efforts. The volume gives some
idea as to the range of topics covered by the sessions at the IGC.
The arrangements for the Congress were very good and all the
participants enjoyed and benefitted from the results of the
meeting, although not on a first-hand basis.

Wichita, Kansas   D. F. Merriam
May 1988

LIST OF CONTRIBUTORS

Abasov, M. T., Azerbaijan Academy of Sciences, 33 Narimanov Avenue, Baku, USSR 370143

Agterberg, F. P., Geological Survey of Canada, 601 Booth Street, Ottawa K1A 0E8, Canada

Bawiec, W. J., U. S. Geological Survey, National Center, Reston, Virginia, 22092, USA

Beinkafner, K. J., P.O. Box 3843, Casper, Wyoming 82602, USA,; present address: 115C Windfield Corners Road, Stone Ridge, New York 12484, USA

Carr, J. R., Deparment of Geological Engineering, University of Missouri, Rolla, Missouri 65401, USA

Chai, Junjie, Institute of Geology, Academia Sinica, Beijing, China

Djafarov, I. S., Azerbaijan Academy of Sciences, 33 Narimanov Avenue, Baku, USSR 370143

Djafarova, N. M., Azerbaijan Academy of Sciences, 33 Narimanov Avenue, Baku, USSR 370143

Djevanshir, R. D., Azerbaijan Academy of Sciences, 33 Narimanov Avenue, Baku, USSR 370143

Drew, L. J., U. S. Geological Survey, National Center, Reston, Virginia, 22092, USA

Fabbri, A. G., Instituto di Geologia Marina, CNR, Via Zamboni, 40127 Bologna, Italy; present address: Canada Center for Remote Sensing, Methodology Section, 1790 Woodward Drive, Ottawa, Ontario
K2C 0P7, Canada

Harris, D. P., Department of Mining and Geological Engineering, College of Mines, University of Arizona, Tucson, Arizona 85721 USA

Jacquemin, P., Centre de Recherches Petrographiques et Geochimiques, CNRS, 54500 Vandceuvre-les-Nance, France

Jewett, D. G., Department of Geology, The Wichita State University, Wichita, Kansas 67208, USA; present address: Wagner, Heindel, & Noyes, Inc., P.O. Box 1629, Burlington, Vermont 0542, USA

Kacewicz, M., Department of Geology, Warsaw University, Warszawa, Poland

Kretz, R., Department of Geology, University of Ottawa, Ottawa, Ontario
K1N 9B4, Canada

Liu, Chenguo, Institute of Geology, Academia Sinica, Beijing, China

Loudon, T. V., British Geological Survey, Edinburgh EH9 2LF, Scotland, UK

Mallet, J. L., ENSG, Ecole National Sureriere de Geology, 54500 Vandceuvre-les-Nancy, France

Merriam, D. F., Center for Applied Geological Research, The Wichita State University, Wichita, Kansas 67208, USA

Royer, J. J., Centre de Recherches Petrographiques et Geochimiques, CNRS, 54500 Vandceuvre-les-Nancy, France

Schuenemeyer, J. H., Department of Mathematical Statistics, University of Delaware, Newark, Delaware 19711, USA

Spease, C., Department of Geological Engineering, University of Missouri, Rolla, Missouri 65401, USA

CONTENTS

SPATIAL MODELING BY COMPUTER[1]

T. V. Loudon

British Geological Survey

ABSTRACT

Computer-based spatial models, representing the past and present disposition and configuration of sets of rock bodies, are a potentially important segment of a geological knowledge base. A spatial model is an interpretation that should be consistent with available data and with expectations based on knowledge of the processes which formed and deformed the rock bodies. The expectations refer not to the structure of the processes, but to their effects, which can be expressed as regional patterns, descriptive statistics, spatial relationships, material budgets, and balances. By using measures of these effects to control and modify interpolation, they can be taken into account in the modeling of surfaces and lines for display, measurement, analysis, and prediction. Processes which create features too small to be located from the data, contribute to an uncertainty envelope which can be defined around the interpolated surfaces.

The conventional method of describing a conceptual spatial model is with geological maps and cross sections. These static two-dimensional images have limitations, not shared by digital models, in representing a fuzzy pattern in three dimensions of

---

[1] Published with permission of the Director of the British Geological Survey (N.E.R.C.).

complex interrelated surfaces. Maps are nevertheless unsurpassed as an aid to retrieval and visualization of spatial information. The advantages of both map and model can be obtained by the user generating displays from a digital model through interactive computer graphics.

SPATIAL MODELING BY COMPUTER

One of the major challenges currently facing a geological survey is the transformation of its geological knowledge base to digital form. The task perhaps is more wide-ranging than in, say, the oil industry (Davis, 1981) because of the diversity of information and generality of objectives in a survey, and progress consequently seems slower. Spatial modeling is an important aspect of the transformation, for it refers to what has long been a core activity of geology, namely piecing together a picture of the geometrical configuration and disposition of rock bodies, their relationships in space, their constituent materials, characteristics and properties, and relating that picture to ideas of their origin and history.

Without the computer, the spatial model exists only in the mind of the geologist. Information about it is communicated through reports and memoirs, maps, and cross sections. Significantly, it is not the model itself that is communicated, but images that have the same relationship to the model that a photograph does to a solid object. The model is about sets of complexly related variables in three-dimensional space and their evolution in time. No static two-dimensional image can depict fully such a model, much less can strings of text. In principle a precise digital representation is possible, allowing the spatial model to be communicated directly, and to become the shared creation and property of the geological community.

The spatial model occupies a gap in current computer applications. On one side of the gap, completed interpretations as embodied in the geological map are becoming available selectively in digital form as computers at last achieve a cost-effective role in the production line for map publication (Boyle, 1980). On the other side of the gap, geological database management now is a well-established activity. However, the observational data available to a geologist typically refer to only a small part of the rocks of interest. Datasets from various sources may have been measured at different points, have imprecise operational definitions, different sampling schemes, and different spatial resolutions.

Thus they are not comparable directly, most methods of data analysis do not apply, and an overall picture can emerge only from skilled and experienced interpretation. Computers can assist in arriving at that interpretation. For example, hypotheses may be suggested and tested using a spatially referenced database linked to contouring algorithms and interactive graphics for selective display of several datasets. With few exceptions, however, the interpretation or model is not generated by computer nor held in digital form. Each user refers back to the original database. Yet for most users, it is the interpretation, not the raw data, which is of primary interest. A digital spatial model could fill the gap if it could be related directly to the database against which the interpretation can be tested and also could be capable of generating the required maps and cross sections as two-dimensional projections of the model. The model should be consistent internally, and consistent with background knowledge and all available data within the limits of their reliability.

The present state of the computing art does not suggest that ideas, interpretations, or background knowledge could come from anywhere other than the geologist. However, digital models could give a better medium to record these systematically, link them with the relevant data, reconcile diverse sources of information, explore the consequences, display the results of the work, and transmit them to the users. Focusing on the model emphasizes the collection of data as raw material for developing, testing, and refining hypotheses, not as an end in itself.

Visualization of the implications of a model must depend on the unsurpassed ability of the human eye and brain to observe and interpret complex graphic images. The geological map is a highly effective communication device. Visual inspection of it can reveal what rocks are present, their distribution, and the spatial relationships between rock bodies and structural and topographic features. Complex information retrieval and an understanding of relationships which defy numerical analysis are achieved readily by visual examination. By linking graphic display techniques to a digital model, eye, brain, and computer can each perform the functions to which they are best suited.

It may be useful to consider the potential benefits that could be expected from a computer model as opposed to conventional methods. A map can give only a limited view of a geological sequence of three-dimensional surfaces and their evolution through time. In the computer model, internal representation is in three or four dimensions throughout, and the functions of data recording, storage and retrieval, analysis and interpretation, and display are handled separately, rather than being combined in one document in a static, two-dimensional format. The results of separation are greater flexibility, higher precision leading to more accurate results, and a more comprehensive record of the geology. A production system also should offer quicker and easier access to the information, lower costs, and savings of manpower. The benefits from the computer model for each separate function are considered in more detail.

In data recording, it is clear that a wide range of quantified information can be supplied to the model. The model, unlike the map, has no defined scale, although it has the comparable attribute of resolution, that is, the distance apart of two points that can be distinguished separately. It is suggested that models should be stored for each area at a number of predetermined levels of resolution. Nevertheless, there is no limit to the fineness of detail, as this can change from one point to another depending on the spatial density of information. The geologist's broad overall impressions also can be quantified. For example, the nature of internal variation in a rock body, such as lateral and vertical changes and gradients, minor discontinuities and repetitive sequences, could be recorded as a summary description rather than showing an artificial homogeneity within formations. Surfaces which can be correlated only locally can be included in a spatial model as well as widespread "mappable units." Work on seismic stratigraphy illustrates their significance (for example, Mitchum, Vail, and Sangree, 1977). The spatial relationships between items can be defined. For example, the relationships that a fossil locality is on a Carboniferous outcrop which is on the downthrown side of a fault, or that a formation boundary converges upstream with a river, can be indicated explicitly, as can the locational accuracy of observations.

Information from the geologist's interpretation or background knowledge can be recorded, and distinguished from observations. The geologist, for example, might have expectations about the general form or shape of a surface from his knowledge of its mode of deposition or structural deformation. On similar grounds, he might expect similarity or accentuation of form between two surfaces, such as the thickness of a sequence of sand bars being greatest at high points on the underlying surface. The expected form of structural surfaces, such as faults or axial planes, also may be known, and geometrical consequences of their interaction with

stratigraphic horizons may be predictable. If the geologist is able
to specify an evolving pattern of changes of form through time
then this too could be included in the model.

Storage and retrieval within a spatial model offer new possibilities.
Alternative explanations or multiple hypotheses could be stored
and the selection and justification of a preferred explanation could
be explicit. The consequences of new data or new hypotheses
could be explored and the model revised to include new
information. As the information is in digital form, it could be
stored for an indefinite time and transmitted rapidly to remote
devices without loss of precision. The items of information to any
level of detail could be identified separately, thus allowing precise
cross referencing within the model and externally from other
models, text descriptions, or data files.

The observations, interpretation, and background knowledge
recorded in the spatial model can each be identified and used in
analysis and display. The analysis can give rise to additional
information which may assist in the interpretation. For example,
geometrical features, such as lines of curvature or small anomalies
on a regional slope, which may be of scientific and economic
interest, can be computed from a digital model and displayed
appropriately, although conventional contouring methods would
not reveal them. The geometrical characteristics of one or more
surfaces, including such aspects as shape, and probable sizes and
numbers of closures, could be summarized numerically from a
model. The similarities of form of successive strata, and the
relationship of biostratigraphic, lithostratigraphic, and seismic
reflector horizons with one another and with faults, axial planes,
and other structural surfaces could be defined and quantified.
The shape descriptors could form the basis of quantitative
background knowledge of regional and vertical variations of form,
and of the influence of mode of deposition and structural
deformation. This in turn could be used explicitly in the
interpretation, with each step explained and justified in a linked
commentary file. At each stage, uncertainty could be estimated
quantitatively and its sources in observation or interpolation
defined.

Derived information, such as areas, volumes, and slope or
curvature distributions could be calculated directly from a digital
model for use in resource estimation, prospectivity analysis,
sediment budget calculations, reconciliation of interpreted
deformation patterns with the theoretical stress field, etc. Some

important sources of data, such as gravity or regional geochemical measurements, cannot define the geometry of the spatial model uniquely. However, the conceptual model can be adjusted to reconcile its predictions with the observed data, within the limits of statistical significance. Thus, there is scope for integrating many diverse sources of information within a model. Derived or implicit relationships (such as: because A is younger than B, and B is younger than C, then A is younger than C) can be computed and any inconsistencies brought to the attention of the geologist for correction.

The computer model can offer a wide variety of forms of output, and remote access to them through the telephone network. Printed output might be required for such results as: prognosis of depths to defined horizons at a proposed well; area or volume calculations; shape descriptors; frequency and orientation distribution of closures of various sizes, etc. Graphic display can be flexible, and such forms as line and symbol maps, posted value maps, contours, isopachs, perspective views, cross sections, fence diagrams, stereograms, correlograms, form surface, and lines of curvature maps are possible outputs. The content, scale, and type of computer-drawn displays could be selected by the user, and it is practical to generate series of diagrams to examine different aspects of the geology. They could be based on the most recent data and interpretation. The constraints and controls arising from the interpretation must be available therefore to the interpolation and display algorithms.

Computer-graphics techniques, including raster display with texture and color, stereo, moving, and perspective views (Foley and van Dam, 1982), can assist in visualizing three-dimensional form, changes through time, levels of uncertainty, and variability within formations. Where there is sufficient compatibility, selected information from the model could be made available to users in digital form, for manipulating or displaying with other digital spatial data, such as mine plans, catchment areas, land-use maps, etc.

Overall, in the face of growing specialization and fragmentation of geology, the computer-based spatial model should help the geologist to bring together information from several disciplines; to explore hypotheses and analyze his information more fully; and to display results appropriately for the user needs of the future. It could help to restore the balance between the large amounts of data potentially available to the geologist, and the limited ability of the geologist to analyze and interpret his data. The model could give greater coherence to computer developments in geology in such areas as digital cartography, text handling, databanks, and graphics by relating them to one core activity.

A project is under way at the British Geological Survey to explore and develop the concepts and strategy in what remains a neglected area of computer application despite its central importance in geology. Years of research, discussion, training, and experience will be required for full implementation, not least because it must rely on the skills of geologists at present unused to computer techniques. The benefits are long term, but conventions once adopted tend to persist, and international collaboration at an early stage may be desirable.

REFERENCES

Boyle, A.R., 1980, Scan digitization of cartographic data, in Freeman, H., and Pieroni, G.G., eds., Map data processing: Academic Press, New York, p. 27-46.

Davis, J.C., 1981, Looking harder and finding less-use of the computer in petroleum exploration, in Merriam, D.F., ed., Computer applications in the Earth sciences, an update of the 70's: Plenum Press, New York, p. 125-144.

Foley, J.D., and van Dam, A., 1982, Fundamentals of interactive computer graphics: Addison-Wesley, Reading, Massachusetts, 664 p.

Mitchum, R.M., Vail, P.R., and Sangree, J.B., 1977, Stratigraphic interpretation of seismic reflection patterns in depositional sequences, in Payton, C.E., ed., Seismic stratigraphy-applications to hydrocarbon exploration: Am. Assoc. Petroleum Geologists Mem. 26, p. 117-133.

METHODS OF THEMATIC MAP COMPARISON

D. F. Merriam and D. G. Jewett

The Wichita State University

ABSTRACT

Geologists are interested in comparing maps in order to (1) determine their predictive value, (2) evaluate their similarity and classify them, and (3) use the information for geological interpretation. The comparisons are made by constructing a difference map (isopachous maps) or on a point-by-point basis (computing an overall correlation coefficient for the fit).

Alternatively, surfaces may be represented by numerical descriptors which can be used as the basis for comparison. The original data points need not be at the same location so that different geographic areas can be compared. However, a uniquely defined spatial grid mesh must be overlain on each data set for the purpose of interpolating grid values which may be compared quantitatively. The interpolated grid values then are compared to determine "reliability indices" at individual grid node locations. This newly generated grid matrix is contoured and shows in two dimensions which areas are most alike and which ones are most dissimilar.

INTRODUCTION

Until recently, maps have been compared mainly by visual inspection. That is, one map is overlain on another and the similarities and dissimilarities noted. If there are only two or three maps to be compared and the pattern is not too complex, this visual comparison is extremely effective - the human mind is excellent in sorting out patterns recorded by the human-eye optic scanner. However, if many maps are involved and the patterns are intricate - the visual comparisons becomes tedious, if not impossible. Also, no two investigators will arrive at the same conclusion on degree of similarity of comparison, unless the patterns reinforce each other and the trends are prominent. To overcome these problems, a series of quantitative techniques have been adapted and developed to automate the comparison procedures to determine areal correspondence.

Geologists, for many years, have used the direct approach of overlaying and subtracting one map from another to obtain a difference or thickness map of an interval. The same effect is obtained by taking data points common to both maps and subtracting the values (of course assuming that both variables are in the same units) and then contouring the data. Several isopachous maps might by prepared for one area depending on the geological problem. These are examples of studies involving several variables in a single area either with data points common to all data sets or different data points but in the same area. There are also other situations where it is desirable to compare the distribution of a particular variable in different areas.

Another combination is to study the spatial distribution of different variables in the same area. For example, it may be desirable to know the spatial relationship of geological, geophysical, or geochemical parameters. It might be of interest to determine the change of a variable through time. For instance the change in chemistry in a body of water or the movement of material on a beach. It may be that it is necessary to determine the relationship of a variable at different locations, thus a problem might be to determine the amount of structural complexity in different parts of a basin or relate the pattern of mineral occurrence in different areas. In each of these situations, there is a procedure to make comparisons.

The result of the quantitative comparisons can either be (1) a resultant map, which shows the areas of greatest similarity and dissimilarity, or (2) a single value noting the 'goodness-of-fit', but does not indicate where. For general comparison of areas, a method to compute a coefficient of areal correspondence can be used (Unwin, 1981). It is expressed as

$$C_a = \frac{\text{area over which phenomena are located together}}{\text{total area covered by the two phenomena}}$$

and there are several variations on this theme which have been developed for different situations. If the parameters are in different units, then the data have to be standardized first and the computed numerical descriptors used for the comparison.

Where the maps being compared are from the same area, it is assumed that the maps have the same scale and the same orientation. Where the maps being compared are from different areas, and thus have no common geographic reference point, the scale must be the same but, obviously there is no orientation. Other problems are, of course, different investigators can arrive at different results; using different approaches to the same problem may give different results; and the comparison method may have no statistical basis - thus no confidence level can be assessed.

PREVIOUS WORK

Although there is an extensive literature on the subject in geographic, image-analysis, and pattern recognition journals, surprising little has been written in geology. A good summary of map comparison can be obtained in Unwin (1981), Gold (1981), and Davis (1986).

Several methods have been used by geologists in comparing maps - each are fairly simple and each have drawbacks. Regression analysis has been used to predict the thickness and number of discrete sandstone beds in a stratigraphic unit (Ribeiro and Merriam, 1979). This technique also has been used to predict the structural configuration of one map to another. Another direct approach is to take corresponding data points from each map and plot them on a scatter diagram to determine the degree of accordance (Mirchink and Bukharsev, 1959). Other techniques involve computing numerical approximations of the surfaces and using these representations for comparison.

Miller (1964) proposed to compare maps with their trend-surface matrices and (Rao, 1971) used the least-square summation equation. Although the grid values are numerical descriptors of the surface, the matrices could be differenced just as could be the original data. Merriam and Sneath (1966) used the coefficients of a low-degree polynominal trend as numerical descriptors of surfaces, and Thrivikramaji and Merriam (1976) used the coefficients from trend surfaces fit to isopachous maps to help interpret structural development of the area. Residuals of the trend surfaces were used by Merriam and Lippert (1964, 1966) in a structural study. Although it has not been done, there is no reason that the coefficients and residuals of Fourier surface fits could not be used also. Robinson and Merriam, (1972), used values at grid positions on spatially filtered maps, and Srivastava and Merriam (1976) determined power-spectra functions have value as numerical descriptors, too. Whitten and Koelling (1973) proposed spline-surface coefficients could be used to compute interpolated values and contours for a map and presumably could be used as numerical descriptors. There undoubtedly are other algebraic and statistical techniques that could be used for describing surfaces numerically.

COMPARISONS

If maps are to be compared where the data set and area are the same, then one set can simply be subtracted from the other and the resultant values contoured; this in essence is an isopachous map. It also is possible to subtract one contoured map from the other if the original data points are not available. If, however, the values to be compared are in different units, it is necessary to first standardize them (Table 1). Likewise if the comparison is to be based on numerical approximators the values have to be standardized first.

In some instances it may be desirable to pretreat the data, that is filter out unwanted components of the data set. This approach has been used successfully by Robinson and Merriam (1984) in their studies of the relationship of oil accumulation to local structure in the Midcontinent.

The maps then are rescaled for ease in computation and the same scaling factor is used on all maps to insure relative amplitude information is obtained, The data matrices (maps) are by normalized using the standard normal forms taking a data value, subtracting the mean, and dividing by the standard deviation. The prepared data sets are multiplied element by element (they could be added or subtracted) and a similarity map computed. Simple cross multiplication produces a new map that is no longer a linear function of the input maps. Similar features of high amplitude are given undue precedence over equally similar features of lower amplitude. Taking the square root of the product values restores

Table 1. Types of variables used in comparisons (from Merriam
          and Robinson, 1980)

data	Variables	
	same units	different.units
original data (data points)	subtract one surface from another [a]	standardize and make direct comparison [b]
numerical approximators	direct comparison [c]	standardize and compare [d]

    (a)   structure on two horizons (same data points) - one area
    (b)   grain size vs porosity (same data points) - one area
    (c)   approximators (e.g. ts) of two horizons - one area
    (d)   approximators of gravity measures vs structure - one area

a form of linearity, and produces a better perspective of spatial
similarity between component features (Merriam and Robinson,
1981).

The resultant maps of the cross-multiplication operation have the
following characteristics. A flat area on one map compared with a
high or low on the other map would give a near-zero value (Table
2). Two corresponding highs or lows would result in a positive
value; coincidence of a high and low would result in a negative
value.

EXAMPLE

Numerous examples of map comparisons can be given but only
one is presented here. An area in south-central Kansas is used as
an example where three structure maps and the topographic map
of the same area were compared. The data were gridded on a
6-mile interval, the values standardized, and cross-multiplied as
noted. The results are given in Table 3.

The coefficients indicate the degree of similarity - high plus
values are positive and low minus values negative. The values
show that the topographic surface is related inversely to the
Precambrian configuration but related positively to the structure
of the Permo-Pennsylvanian units. The Precambrian also is
related inversely to the structure on the Permo-Pennsylvanian and
Mississippian horizons. The near-zero values for the
Mississippian structure in relation to the topography and
Permo-Pennsylvanian structure are not significant geologically.

Table 2. Map features resulting from multiplication (from Merriam and Robinson, 1981)

	Resultant Map Feature		
	positive	zero	negative
comparison function times (x)	⌢⌣	⌢—	⌢⌣
	⌣⌣	—⌣	⌣⌢

These single values give some insight into the overall relation of one map to another but do not indicate how or where the coincidence occurs. Therefore, it is not possible to tell if the maps are roughly similar over the entire surface or whether there is a high coincidence locally but little similarity in other places.

RELIABILITY INDICIES

There are numerous methods by which reliability indices can be determined for pairwise compared maps. One way is to take the absolute difference of the standardized grid matrix value at a point common to both maps. As the index approaches zero, the two matrix values are more similar and as the index approaches infinity, the two values are less similar. If these indices are calculated for each grid node, then the values can be contoured and a map is created showing the areas most and least alike in configuration.

Another method, and the one we have selected to use, is a two-dimensional smoothing function. A correlation coefficient for a grid-node value and surrounding square of eight values is computed, the square then is moved to the next grid node and another correlation coefficient computed for those nine values, the square is moved to the next grid mode and a coefficient computed, etc., until the entire map has been covered. These correlation coefficients, excluding the bordering values of the map which are lost due to edge effects then are contoured. The resultant map shows plainly where the configuration of the pairwise comparison coincides and where it does not. Where the

Table 3. Degree of correspondence of comparison of pairs of
          surfaces

TOPO	–			
P-P	0.84	–		
MISS	0.13	0.09	–	
PЄ	-0.77	-0.92	-0.92	–
	TOPO	P-P	MISS	PЄ

maps coincide, then, the reliability index will be high, and where
it is low the coincidence will be low; predictability will be good
where the reliability index is high.

SUMMARY

The results of a map comparison can be: (1) a difference map, or
(2) a "goodness-of-fit" coefficient. The difference map may be
obtained by: (1) subtracting values at the original data points, or
(2) standardizing the data and subtracting. The advantage of a
difference map is that it will show areas where correspondence is
good (or poor) but will not give the degree of similarity. The
advantage of a coefficient is that the data points need not be at the
same location, in fact, even different areas can be compared (if
the scale is the same). Again, this will show the degree of
similarity but not the location. A reliability index may be
computed to determine the areas on the resultant map which are
most and least alike in configuration.

ACKNOWLEDGMENTS

We would like to thank our colleagues John E. Nordstrom and
Gerard V. Wolf for help in preparing the original data set and for
comments and suggestions. They also helped in preparing
material for the presentation at the 27th International Geological
Congress in Moscow 1984. Revisions of the original paper were
made for a presentation to the South-Central Section meeting of
the Geological Society of America in 1985; and it was given again
at a NATO ASI conference on "Statistical Treatment for

Estimation of Mineral and Energy Resources" held at Il Ciocco
(Lucca) Italy in June of 1986. Linda Spurrier kindly typed the
manuscript and Mark Sondergard helped with the special
programming.

REFERENCES

Davis, J.C., 1986, Statistics and data analysis in geology (2nd ed.):
John Wiley & Sons, New York, 646 p.

Eschner, T.R., Robinson, J.E., and Merriam, D.F., 1979,
Comparison of spatially filtered geologic maps: summary:
Geol. Soc. America Bull., pt. 1, v. 90, p. 6-7.

Gold, C., 1980, Geological mapping by computer, in Taylor, D.R.F.,
The computer in contemporary cartography: John Wiley &
Sons, Chichester, p. 151-190.

Merriam, D.F., and Jewett, D.G., 1985, Quantitative comparison of
thematic maps (abst.): Geol. Soc. America Abstracts with
Programs, 19th Ann. Meeting South-Central Sec., v. 17, no.
3, p. 167.

Merriam, D.F., Jewett, D.G., Nordstrom, J.E., and Wolf, G.V., 1984,
Methods of thematic map comparison (abst.): 27th Intern
Geol. Congress (Moscow), Abstracts v.8, secs. 17 to 22,
p. 389.

Merriam, D.F., and Lippert, R.H., 1964, Pattern recognition
studies of geologic structure using trend-surface analysis:
Colorado Sch. Mines Quart., v. 59, no. 4, p. 237-245.

Merriam, D.F., and Lippert, R.H., 1966, Geologic model studies
using trend-surface analysis: Jour, Geology, v. 74, no. 3,
p. 344-357.

Merriam, D.F., and Robinson, J.E., 1980, Numerical description,
enhancement, segmentation and comparison of thematic
maps: Sciences de la Terre, Informatique Geologique,
no. 15, p. 11-29.

Merriam, D.F., and Robinson, J.E., 1981 Comparison functions and
geological structure maps, in Future trends in
geomathematics: Pion. Ltd.,London, p. 254-264.

Merriam, D.F., and Sneath, P.H.A., 1966, Quantitative comparison
of contour maps: Jour. Geophysical Res., v. 71, no. 4,
p. 1105-1115.

Miller, R.L., 1964, Comparison-analysis of trend maps, in
Computers in the mineral industries, pt. 2: Stanford Univ.
Publ., Geol. Sci., v. 9, no. 2, p. 669-685.

Mirchink, M.F., and Bukhartsev, V.P., 1959, The possibility of a
statistical study of structural correlations: Doklady Akad.
Nauk SSSR (English translation), v. 126, no. 5,
p.1062-1065.

Rao, S.V.L.N., 1971, Correlations between regression surfaces
based on direct comparison of matrices: Modern Geology,
v. 2, no. 3, p. 173-177.

Ribeiro, J.C., and Merriam, D.F., 1979, Quantitative analysis of
depositional environments (Aratu Unit, Reconcavo Series,
Lower Cretaceous) in the Reconcavo Basin, Bahia, Brazil, in
Geomathematical and petrophysical studies in
sedimentology: Pergamon Press, Oxford, p. 219-234.

Robinson, J.E., and Merriam, D.F., 1972, Enhancement of
patterns in geologic data by spatial filtering: Jour. Geology,
v. 80, no. 3, p. 333-345.

Robinson, J.E., and Merriam, D.F., 1984, Computer evaluation of
prospective petroleum areas: The Oil and Gas Jour., v. 82,
no. 34, p. 135-138.

Srivastava, G.S., and Merriam, D.F., 1976, Computer constructed
optical-rose diagrams: Computers & Geoscience, v. 1, no. 3,
p. 179-186.

Thrivikramaji, K.P., and Merriam, D.F., 1976, Trend analysis of
sedimentary thickness data: the Pennsylvanian of Kansas, an
example, in Quantitative techniques for the analysis of
sediments: Pergamon Press, Oxford, p.11-21.

Unwin, D., 1981, Introductory spatial analysis: Methuen, London
and New York, 212 p.

Whitten, E.H.T., and Koelling, M.E.V., 1973, Spline-surface
     interpretation, spatial filtering, and trend-surfaces for
     geological mapped variables: Jour. Math. Geology, v.5, no. 2,
     p. 111-126.

A STUDY OF TWO-DIMENSIONAL GRAIN SEQUENCES IN ROCKS

Andrea G. Fabbri and Ralph Kretz

University of Bologna and University of Ottawa

ABSTRACT

The visual aspect of crystalline fabrics in thin or polished sections under the microscope is captured by digitizing the outlines of all recognizable grain profiles (from 250 to 2500) over areas of 3 to 6 cm^2. Digital image processing of the textural data obtained is applied to extract quantitative aspects not detected readily by human vision such as the frequency distribution of contacts between grains belonging to the same or to different phases. The methods described are of importance when a model of crystallization can be expressed in terms of geometrical relationships between grains. Furthermore, such relationships correspond to physical characteristics of a rock.

This contribution reviews recent work in texture analysis and proposes some new techniques in which computer processing and petrology are useful mutually for quantitative measurement and recognition. Examples of applications to metamorphic rocks and intrusive rocks are described in which image processing leads to the following results: (a) the probability of a grain to be surrounded by other grains, (b) the mapping of particular grain sequences, and (c) the identification and statistical analysis of individual grain contacts.

INTRODUCTION

The three-dimensional arrangement of grains in polycrystalline aggregates is a reflection of both their mode of crystallization and their physical properties. Additionally, deformation and recrystallization may add their imprint to the aggregate, possibly obliterating, partly or fully, previous arrangements. In a broad sense, the term texture can be referred to all geometrical characteristics of the grains, and distribution in space. Because grains of the same type (or phase) may tend to cluster within an aggregate, the shape and distribution of those clusters also are textural properties. Therefore, in textural studies it is necessary to define some primitive components (grain, or group of grains) and to characterize their morphology, spatial arrangement, and interrelations.

Three-dimensional samples, such as those used by Kretz (1966) rarely are obtainable in practice, and rocks are studied from thin or polished sections from which the apparent morphology and arrangement of grains can be observed. Quantitative characterization of textures cannot be done easily visually, but requires particular procedures for digitization and computer processing.

A recent study by Vistelius and others (1983) exemplifies how properly spaced parallel linear traverses across thin sections of granitic rocks can be used to code the occurrence of various crystalline phases and to compute matrices of transitions for microcline, quartz, and albite. Different patterns of transitions have been related to differences in viscosity and in content of volatile components.

Kretz (1969) probably was the first to give particular attention to the statistical analysis of grain profiles for quantitatively relating their geometrical attributes to nucleation and crystallization processes that could be modeled for about 1000 grains manually traced from a thin section of pyroxene-scapolite-sphene granulite. At the level of detail selected, the rock was determined to be homogeneous and the three crystal phases seemed to be distributed randomly in the section.

A review of studies of rock textures for the analysis of microscopic thin sections has been made by Fabbri (1984), who proposed to use digital image processing to quantify and study such textural information. The visual aspects of crystalline fabrics is captured by digitizing the outlines of all recognizable grain profiles (from 250 to 2500) over areas of 3 to 6 cm^2. Processing of the data obtained can be applied to extract quantitative aspects not detected readily by human vision, such as the frequency and distribution of contacts between grains of the same or of different phases.

This paper considers recent work in texture analysis and proposes some techniques in which computer processing and petrology mutually are useful for quantitative measurement and recognition. Three examples of application to metamorphic and intrusive rocks lead to new results.

CRYSTALLIZATION

Crystallization, either from a silicate melt (igneous crystallization) or in solid rock (metamorphic crystallization) is a naturally occurring chemical process. The arrangement of atoms to produce a crystal always occurs with a release of Gibbs free energy; the entropy of the system may increase or decrease but total entropy (system + surroundings) invariably increases. In general, crystallization in igneous rocks is induced by a drop in temperature, and in metamorphic rocks by a rise in temperature; changes in pressure also may play a role.

The creation of a crystal may be viewed as occurring in two steps. Initially, a nucleus if formed (nucleation) and then the crystal becomes larger in size (crystal growth). Once formed, crystals also may decrease in size, that is experience negative growth. Usually, during the early stages of growth, crystal faces are present; these may disappear as crystals impinge on each other.

Questions concerning the size, orientation, and spatial distribution of crystals in rocks may be examined in relation to nuculeation. The classical theory of nucleation supposes that as a system is displaced slowly from the point of equilibrium, embryos (with a surface energy exceeding the gain in bulk free energy) are built and demolished repeatedly, until an embryo reaches the critical size (where surface energy becomes subordinate to bulk free energy) and becomes a nucleus. this theory may be modified and applied to mineral crystallization in rocks.

With regard to solid-state crystallization in metamorphic rocks, the theory postulates the presence of countless possible nucleation sites and numerous different types of sites, requiring different amounts of interfacial energy to accommodate a nucleus. The rate at which nucleation sites are occupied may be important in determining the size of crystals; slow nucleation (relative to the rate of reaction) would produce few large crystals, whereas rapid nucleation would produce many small crystals. With some sites energetically more favorable than others, a crystal-size distribution would be obtained, from which inferences may be drawn regarding the rate of nucleation. When crystallization occurs in a stress field, the strain energy of embryos may play a role, and embryo with a certain orientation more readily may become nuclei, resulting in a preferred orientation of crystals. Nucleation also may determine the spatial distribution of crystals; if favorable nucleation sites are clustered, the crystals that occupy these sites also will be clustered.

In igneous rocks, particularity in gabbroic dikes, the rate of cooling (which may be estimated) obviously is important in determining nucleation rates, and hence crystal size. The dramatic decrease in the size of crystals as the contact is approached is the consequence of a large increase in the rate of nucleation, resulting from an increase in the rate of cooling. This is in agreement with the classical theory, which predicts that the critical nucleus size is smaller in dike margins, where the bulk free energy term would be large. Within certain limits, different minerals may experience different rates of nucleation at a given rate of cooling, causing for example small crystals of one mineral to be enclosed partially by larger crystals of another mineral (subophitic texture). In some rocks, certain crystals have nucleated evidently upon the surface of another crystal, and if these are of the same type, crystal aggregates will be produced, for example, aggregates of augite in gabbro.

Numerous possibilities exist for the growth of a crystal, with regard to both mechanisms and rates. The general presence of crystal faces indicates that growth occurs by some mechanism involving the arrangement of atoms on crystal faces as the rate-determining process. The rate of increase of the radius of a crystal is constant, provided the rate of increase of volume is proportional to surface areas. A growth law of this type is suggested by some zoned crystals.

In metamorphic rocks, a growing crystal must displace minerals lying in the path of growth; this displacement may be physical or chemical in nature. Where the displacement is not successful entirely (resulting possibly from rapid growth) various patterns of inclusions may result, for example quartz crystals.

An astonishing diversity exists in the texture of crystalline rocks, and new methods of obtaining and quantifying textural data are needed urgently, as well as new interpretations of such data, to increase our understanding of rock-forming processes. It seems likely at present that nucleation (and the effect of temperature and strain on nucleation) is of great importance in determining the textural characteristics of many rocks.

MODELS OF CRYSTALLIZATION

A concept of an ideal magmatic granite was proposed by Vistelius (1972) for theoretical situations in which crystallization from a melt occurs at constant pressure and in the absence of movement of crystals, while new centers of crystallization slowly separate at each crystallization stage. The model considered "pretectic," "cotectic," and "eutectic" schemes for the system plagioclase-quartz-potassium feldspar, in which ideally crystallization begins for crystal generation of only one mineral, followed by two minerals, and completed by the generation of three minerals simultaneously. A statistical model defines measures of probability of transitions for one phase to another phase by Boolean algebra. For example, for an ideal granite to be considered as a standard for comparison, we would expect sequences of transitions close to "simple (first-order), homogeneous, reversible, ergodic Markovian sequence" (Vistelius, 1972, p. 95). A first-order Markovian property for the sequence is one in which the probability of occurrence for a given phase is due partly and significantly to the phase immediately preceding it along a linear traverse. Deviations from the ideal model represented by a first-order Markovian sequence without some of those additional properties, a second-order Markovian sequence, or a Bernoullian sequence, for example, can be interpreted as due to metasomatic changes or to disturbances that displaced crystalline particles relatively to one another.

More recently, Vistelius and others (1983) used a simplified model in which the sequence microcline-quartz-albite is considered, but the crystallization stages are ignored. This model was justified by the observation that the grains analyzed did not show traces of interruption in crystal growth. The stochastic model applied to a stock of fine-grained potassic granite was analyzed carefully in detail and verified. Statistical tests and a computer program were provided for its application.

Kretz (1969) used a simple model of nucleation and growth in the analysis of a calcium-rich granulite (consisting of calcic pyroxene, scapolite, sphene, calcium amphibole, apatite, and zircon) in which crystals of sphene, pyroxene, and scapolite nucleated at random and the nucleation sites of crystals was not determined or influenced by any other crystal in its neighborhood. This situation could be simulated in two dimensions by generating randomly distributed nucleation sites of black and white grains, then expanded at uniform rates in all directions until complete impingement is obtained.

Flynn (1969) showed that grains in gneisses are arranged spatially so that grains of the same phase tend not to occur in contact with each other; an arrangement interpreted to arise from grain-boundary migration leading to the insertion of one phase between pairs of like grains of another phase. Interfacial energies of contact between like phases may have been greater that those between unlike phases. For this reason, in some metamorphic rocks unlike contacts are favored statistically to like contacts at high level of significance.

Ehrlich and others (1972) observed that surface free energy, which is a function of surface area and grain neighborhoods, may be a dominant factor in petrogenesis. In a study of the response of textural variables to metamorphic grade, they determined progressive increase in the proportion of unlike grains-to-surface free energy. Similarly, an increase in average size of grains, by reducing surface area, also should result in a decrease of surface energy. A rock of a given composition affected by increasing metamorphism only can conserve or decrease its surface free energy by textural readjustment through a limited range. When this range is exceeded, chemical readjustment can be triggered, producing more stable assemblages.

They demonstrated that, in the granodioritic rock studied, adjustments in free energy arose from three principal mechanisms: the growth of grains, change in neighborhoods, and nucleation of new grains. Such textures can be described using concepts of geometric probability and topology. Those authors suggested that the grain boundary network of a rock can be considered as an integrated response of a specific compositional assemblage to petrogenetic processes. Interpenetrating subnetworks can be defined on the bases of selected phase-to-phase contacts.

Jen (1975) determined that, in a study of charnockitic granulites, three major fundamental types of spatial distribution may occur in a rock: clustered, regular, and random (also intermediate types such as antiregular and anticlustered). Because this occurs, although in most of the rocks there are two instead of all three types, such a mixing mode of spatial distribution would be expected to be the usual situation rather than the exception. This suggests that an ideal situation of crystallization in natural rocks is rare, and that total equilibrium is achieved rarely.

Whitten, Dacey, and Thompson (1975) studied sequences of mineral grain transitions along linear transverses across mutually perpendicular sets of serial sections in another sample of the same calc-silicate granulite studied by Kretz (1969). Rather exhaustive tests that they performed proved that the granulite possesses a Markovian property in which observed mineral grains are controlled by the composition of adjacent grains in a rock. Such property, however, seems to reflect petrogenetically important but unidentified factors. In the situation of the calc-silicate granulite, which originally was a sedimentary rock, the Markovian properties must be due to a mineralogy produced by high-grade recrystallization in the solid state. A conclusion that would broaden the applicability of the Markov chain models.

The models described here, suggest that textural characterization may be important for both the study of modes of crystallization and for relating textural and physical properties: they are likely to be of practical economic importance.

DIGITIZATION OF SECTIONS OF ROCKS AND SUBSEQUENT PROCESSING

For the applications described in the next section, two methods were used to capture the information from the line drawings of the grain boundaries of the microscopic images of thin sections:

(1) transformation into digital images by scanning of the boundaries of the profiles and computation of a binary (black and white) image of the boundaries, so that the lines are represented by chains of 1-valued pixels (picture elements) of one pixel in width (thin boundary lines); or

    (2) digitization of the boundaries by a graphic tablet (x-y
       digitizer), conversion of the boundary data into a list of x-y
       coordinates in the computer, and conversion of the
       coordinates to a binary image of thinned boundary lines.

Figure 1 shows the end result of either method for a small portion
of a larger binary image of a granitic rock. During subsequent
processing steps, all 0-valued pixels belonging to a grain profile
are assigned automatically a unique numerical label (component
labeling), and all the pixels in grain profiles belonging to the same
crystal type (phase) are assigned interactively a new label (phase
labeling). A phase-labeled image represents an informative
databank from which different textural attributes can be measured
directly, or different binary images can be computed, for example,
one for each phase (binary images of phases for labels 3, 4, and 6).

Additionally, from the binary image of the boundary shown in
Figure 1, both segment and junctions can be labeled automatically
(segment and junction labeled image, see Fig. 2) by unique
numerical labels of positive and negative values, respectively.

Because of the variability in optical characteristics of the mineral
grains in a thin section (color, pleochroism, dishomogeneity in
texture within grains, random-cut effect of grains that show
different thickness at some edges, etc.) it is not possible or
feasible yet for automatic scanning devices to capture and process
sufficient information on large numbers of grains for a satisfactory
phase recognition and extraction. Processing of several optically
scanned images obtained in good registration (pixel-to-pixel
correspondence) for different microscope settings, can be a task
comparable to the pixel classification and feature extraction
employed for remotely sensed data. Therefore, it may not be
justifiable by the economic value of the material studied. For
these reasons, the applications described in this paper are limited
to digitized line drawings produced by hand after meticulous
microscopic study by a petrographer.

However, new techniques are being developed by Reid and others
(1984) and Miller, Reid, and Zuiderwyk (1982) that use a
QEM*SEM image analyzer (Quantitative Evaluation of Minerals by
Scanning Electron Microscopy). This fully automated
instrumentation can determine the type of phases present in a
polished section by an energy dispersive (multielement) X-ray
detector and back-scattered electron detector. A motor-driven

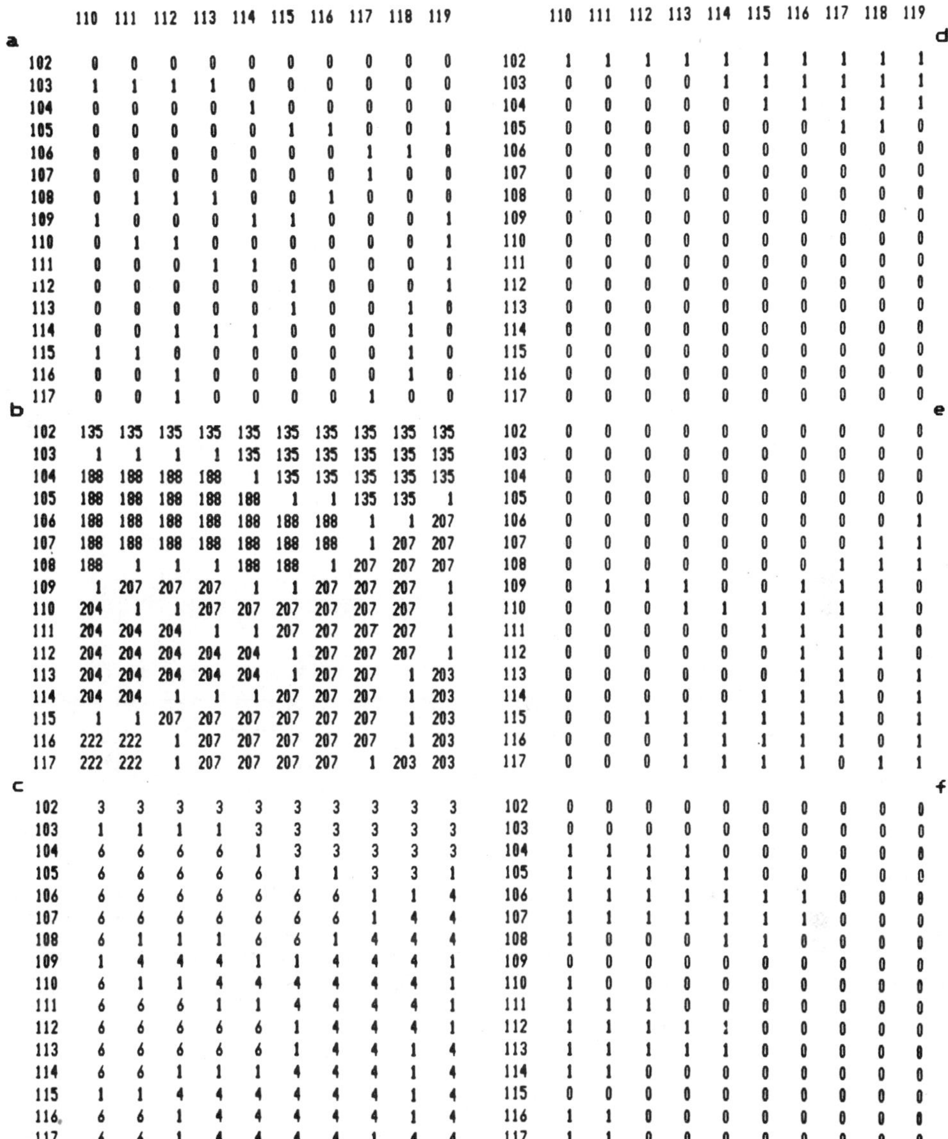

a

	110	111	112	113	114	115	116	117	118	119
102	0	0	0	0	0	0	0	0	0	0
103	1	1	1	1	0	0	0	0	0	0
104	0	0	0	0	1	0	0	0	0	0
105	0	0	0	0	0	1	1	0	0	1
106	0	0	0	0	0	0	0	1	1	0
107	0	0	0	0	0	0	0	1	0	0
108	0	1	1	1	0	0	1	0	0	0
109	1	0	0	0	1	1	0	0	0	1
110	0	1	1	0	0	0	0	0	0	1
111	0	0	0	1	1	0	0	0	0	1
112	0	0	0	0	0	1	0	0	0	1
113	0	0	0	0	0	1	0	0	1	0
114	0	0	1	1	1	0	0	0	1	0
115	1	1	0	0	0	0	0	0	1	0
116	0	0	1	0	0	0	0	0	1	0
117	0	0	1	0	0	0	0	1	0	0

b

	110	111	112	113	114	115	116	117	118	119
102	135	135	135	135	135	135	135	135	135	135
103	1	1	1	1	135	135	135	135	135	135
104	188	188	188	188	1	135	135	135	135	135
105	188	188	188	188	188	1	1	135	135	1
106	188	188	188	188	188	188	188	1	1	207
107	188	188	188	188	188	188	188	1	207	207
108	188	1	1	1	188	188	1	207	207	207
109	1	207	207	207	1	1	207	207	207	1
110	204	1	1	207	207	207	207	207	207	1
111	204	204	204	1	1	207	207	207	207	1
112	204	204	204	204	204	1	207	207	207	1
113	204	204	204	204	204	1	207	207	1	203
114	204	204	1	1	1	207	207	207	1	203
115	1	1	207	207	207	207	207	207	1	203
116	222	222	1	207	207	207	207	207	1	203
117	222	222	1	207	207	207	207	1	203	203

c

	110	111	112	113	114	115	116	117	118	119
102	3	3	3	3	3	3	3	3	3	3
103	1	1	1	1	3	3	3	3	3	3
104	6	6	6	6	1	3	3	3	3	3
105	6	6	6	6	6	1	1	3	3	1
106	6	6	6	6	6	6	6	1	1	4
107	6	6	6	6	6	6	6	1	4	4
108	6	1	1	1	6	6	1	4	4	4
109	1	4	4	4	1	1	4	4	4	1
110	6	1	1	4	4	4	4	4	4	1
111	6	6	6	1	1	4	4	4	4	1
112	6	6	6	6	6	1	4	4	4	1
113	6	6	6	6	6	1	4	4	1	4
114	6	6	1	1	1	4	4	4	1	4
115	1	1	4	4	4	4	4	4	1	4
116	6	6	1	4	4	4	4	4	1	4
117	6	6	1	4	4	4	4	1	4	4

d

	110	111	112	113	114	115	116	117	118	119
102	1	1	1	1	1	1	1	1	1	1
103	0	0	0	0	1	1	1	1	1	1
104	0	0	0	0	0	1	1	1	1	1
105	0	0	0	0	0	0	0	1	1	0
106	0	0	0	0	0	0	0	0	0	0
107	0	0	0	0	0	0	0	0	0	0
108	0	0	0	0	0	0	0	0	0	0
109	0	0	0	0	0	0	0	0	0	0
110	0	0	0	0	0	0	0	0	0	0
111	0	0	0	0	0	0	0	0	0	0
112	0	0	0	0	0	0	0	0	0	0
113	0	0	0	0	0	0	0	0	0	0
114	0	0	0	0	0	0	0	0	0	0
115	0	0	0	0	0	0	0	0	0	0
116	0	0	0	0	0	0	0	0	0	0
117	0	0	0	0	0	0	0	0	0	0

e

	110	111	112	113	114	115	116	117	118	119
102	0	0	0	0	0	0	0	0	0	0
103	0	0	0	0	0	0	0	0	0	0
104	0	0	0	0	0	0	0	0	0	0
105	0	0	0	0	0	0	0	0	0	0
106	0	0	0	0	0	0	0	0	0	1
107	0	0	0	0	0	0	0	0	1	1
108	0	0	0	0	0	0	0	1	1	1
109	0	1	1	1	0	0	1	1	1	0
110	0	0	0	1	1	1	1	1	1	0
111	0	0	0	0	0	1	1	1	1	0
112	0	0	0	0	0	0	1	1	1	0
113	0	0	0	0	0	0	1	1	0	1
114	0	0	0	0	0	1	1	1	0	1
115	0	0	1	1	1	1	1	1	0	1
116	0	0	0	1	1	1	1	1	0	1
117	0	0	0	1	1	1	1	0	1	1

f

	110	111	112	113	114	115	116	117	118	119
102	0	0	0	0	0	0	0	0	0	0
103	0	0	0	0	0	0	0	0	0	0
104	1	1	1	1	0	0	0	0	0	0
105	1	1	1	1	1	0	0	0	0	0
106	1	1	1	1	1	1	1	0	0	0
107	1	1	1	1	1	1	1	0	0	0
108	1	0	0	0	1	1	0	0	0	0
109	0	0	0	0	0	0	0	0	0	0
110	1	0	0	0	0	0	0	0	0	0
111	1	1	1	0	0	0	0	0	0	0
112	1	1	1	1	1	0	0	0	0	0
113	1	1	1	1	1	0	0	0	0	0
114	1	1	0	0	0	0	0	0	0	0
115	0	0	0	0	0	0	0	0	0	0
116	1	1	0	0	0	0	0	0	0	0
117	1	1	0	0	0	0	0	0	0	0

Figure 1: Digitization and processing of grain-boundary data. (a) binary image of thinned boundary lines; (b) component-labeled image; (c) phase-labeled image; (d) to (f) binary images of phases with labels 3, 4, and 6 in (c). Illustrations are for pixel values for subimage of 10 X 16 pixels: above and to left of pixel values are image coordinates for columns and rows, respectively.

	110	111	112	113	114	115	116	117	118	119
102	0	0	0	0	0	0	0	0	0	0
103	3	3	3	3	0	0	0	0	0	0
104	0	0	0	0	3	0	0	0	0	0
105	0	0	0	0	0	3	-8	0	0	4
106	0	0	0	0	0	0	0	-8	-8	0
107	0	0	0	0	0	0	0	-8	0	0
108	0	-12	5	5	0	0	5	0	0	0
109	-12	0	0	0	5	5	0	0	0	7
110	0	-12	9	0	0	0	0	0	0	7
111	0	0	0	9	9	0	0	0	0	7
112	0	0	0	0	0	9	0	0	0	7
113	0	0	0	0	0	9	0	0	7	0
114	0	0	-17	9	9	0	0	0	7	0
115	-17	-17	0	0	0	0	0	0	7	0
116	0	0	-17	0	0	0	0	0	7	0
117	0	0	10	0	0	0	0	7	0	0

Figure 2:   Example of sequential segment-and-junction-labeled image that could be computed from binary image shown in Figure 1a.

moving stage brings successive fields into position, and at each sample point, at regularly spaced intervals, a signal generated by the back-scattered electrons is used to determine the average atomic number of the small area of material irradiated by the beam. "Mineral maps" can be computed, within 1-2 hours, in a digital form that is similar to the one of phase-labeled images.

Although instruments of this type are likely to make the capturing of textural data more accessible in the future, quantitative textural characterization is nevertheless, a complex problem even in the ideal situation of easily available digital data.

EXAMPLES OF PROCESSING TEXTURAL DATA FOR SEQUENCES OF GRAINS

The applications described here exemplify several textural models that can be translated easily into statistical models.

Application #1: Quantitative Characterization of Grain Neighborhoods

In this application, and in the next one as well, the digital image is used that is shown in Plate 1 at the top. It was obtained from a line drawing made by Kretz (1969) who examined in detail the distribution of crystal and crystal boundaries in a thin section of pyroxene-scapolite-sphene granulite of Precambrian age from the Grenville Province of the Canadian Shield. Binary images of the different phases, component-labeled images, and phase-labeled images (similar to the ones shown in Fig. 1) all consisting of 1000x595 pixels are required for the computations that are illustrated in Figure 3.

A set of one or more grains ("cluster seeds") is selected from the binary image of a crystalline phase. If we consider the entire set of binary images that can be extracted from the phase-labeled image (shown in Plate 1 at the top), each grain profile is surrounded by a thin line of boundary pixels. A dilatation (a pixel neighborhood transformation) is computed so that the profiles expand one pixel past the boundary. The dilatated profiles of the selected grains provide the "pointers" to the neighboring profiles that in turn are fetched from the component-labeled image and merged into clusters in a new binary image. This procedure is iterated for the newly constructed clusters, keeping track of the component labels, the phase labels and the numbers of pixels pointed at, for each step of adjacency, for each component, and for each phase. Adjacency relationships are tabulated to provide the basis for testing spatial models of crystallization. Table 1 shows an example for such an adjacency relationship computed for the transformations shown in Figure 3.

By this method of processing, for each grain of a set of interest, tables of adjacencies such as the one illustrated in Table 1, are computed easily. Neighborhoods are characterized by using the number of pixels in each successive "shell", and the frequency of occurrence of different phase labels in each shell. We can think of a distance relationship with each grain, in which the closer we are to a grain, the greater is the dependence of the phases and components encountered moving away from that grain. The farther away we are from the seed (grain or grains) the closer we should get to an average value of grain contact distribution which is typical of the section (average frequency of transitions for all pairs of phases).

The approach presented here can be adapted to different situations, in which the process spans all directions or it advances in one or more specific directions. Markovian properties can be estimated for each set of grains along outward two-dimensional paths. Programming aspects related to this approach have been studied by Fabbri and Kasvand (1981).

Plate 1. Samples represented by the phase-labeled images.

TOP ILLUSTRATION

Applicon color plot of phase-labeled image of granulite digitized from draft by Dretz (1969). Image dimensions are 1000 x 595 pixels. Each pixel corresponds to 1/53 mm or 0.019 mmm and is represented by 2 x 2 array of color dots. There are 1294 grains in this image (within parentheses is number of grains for each phase): red is pyroxene (470), purple is scapolite (574), dark blue is apatite (22), blue is hornblende (61), green is sphene (164), white is zircon (3), black are grain boundary profiles, and light blue is part of image between edges of line drawing and edges of the image, here termed frame.

CENTER ILLUSTRATION

Photograph from high-precision screen of plot of phase-labeled image of granite, digitized from illustration by Kamineni and Katsube (1982). Image dimensions are 420 x 234 pixels; 40 pixels correspond to a distance of 2 mm, i.e. one pixel corresponds to 0.05 mm. There are 201 grains in this image (within parentheses is number of grains for each phase): yellow is K-feldspar (12), green is plagioclase feldspar (34), red is quartz (142), and brown is biotite (13). Permeability for this rock is 20.1 microdarcies (under confined pressure) and 4.2 microdarcies (under unconfined in situ pressure).

BOTTOM ILLUSTRATION

Photograph from high-precision screen of plot of phase-labeled image of granite, digitized from illustration by Kamineni and Katsube (1982). Image dimensions are 371 x 353 pixels; 40 pixels correspond to distance of 2 mm, i.e. one pixel corresponds to 0.05 mm. There are 130 grains in this image (within parentheses is number of grains for each phase): yellow is K-feldspar (20) , green is plagioclase feldspar (38), red is quartz (39), brown is biotite (22), and blue are other phases (19). Permeability for this rock is 0.3 microdarcies (under confined pressure) and 0.12 microdarcies (under unconfined in situ pressure).

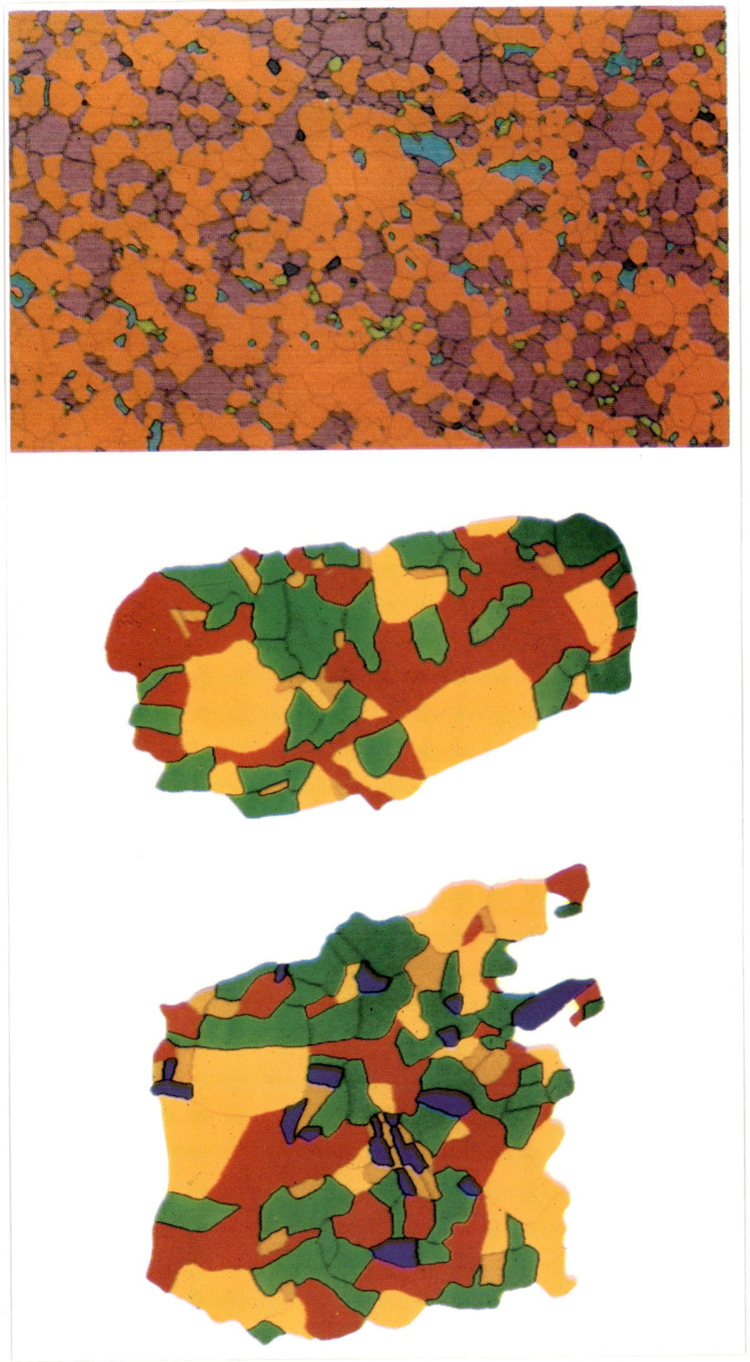

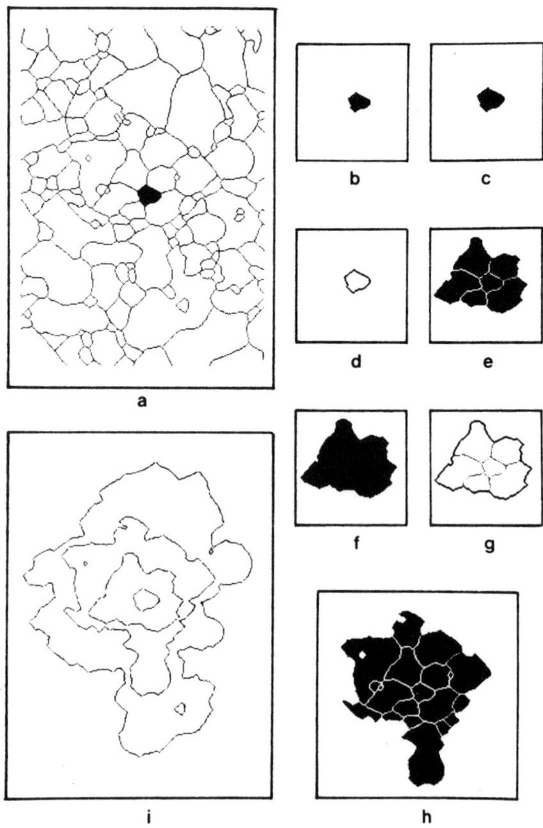

Figure 3. Example of iterative processing of binary image with
one profile of apatite, for computing table of one-step to
four-step adjacency relationships. (a) logical union of
image of selected apatite profile and image of thinned
boundaries; (b) initial seed; (c) dilated seed; (d) black
pixels "added" by dilation of seed; (e) first cluster of
profiles; (f) dilated first cluster; (g) black pixels "added"
by dilation of first cluster of profiles; (h) second cluster
of profiles; (i) image of logical union of four "shells" of
pixels, pointers to one-step to four-step adjacent
profiles: each shell is set of thin-line segments,
interrupted in coincidence with junction-boundary
pixels. Each shell represents probability of adjacency.

Table 1. Porportion of adjacencies between profile of apatite and
its one-step to three-step neighbors in image of
granulite. In brackets coincident pixel counts for each
phase; superscript numbers are numbers of profiles
contributing to count. In upper right corners of boxes
types of adjacency are indicated between phases
occurring in neighborhood of apatite profile: apatite (6),
calcic pyroxene (3), scapolite (4), sphene (5), and
hornblende (7).

```
+--------------+--------------+--------------+
!1-step        !2-step        !3-step        !
!              !              !              !
!              !              !              !
+--------------+--------------+--------------+
!          6 3 !       6 . 3  !    6 . . 3   !
!            1 !            8  !           8  !
!.2159 ( 19)   !.7230 (355)   !.2411 (312)   !
+--------------+--------------+--------------+
!          6 4 !       6 . 4  !    6 . . 4   !
!            3 !            2  !           14 !
!.6023 ( 53)   !.1181 ( 58)   !.5819 (753)   !
+--------------+--------------+--------------+
!          6 5 !       6 . 5  !    6 . . 5   !
!            1 !            3  !           9  !
!.1818 ( 16)   !.0712 ( 35)   !.1569 (203)   !
+--------------+--------------+--------------+
!              !       6 . 7  !              !
!              !            1  !              !
!              !.0876 ( 43)   !              !
+--------------+--------------+--------------+
!Totals      5 !           14 !           32 !
!       ( 88)  !      (491)   !     (1294)   !
+--------------+--------------+--------------+
```

Application #2: Extraction of Two-Dimensional Phase Sequences
from Adjacency Relationships Between Phases

This application represents a modification and generalization of
the approach used in the previous example in that the dilatated
binary image of a phase can be used to point at the grain profiles
of one or more of preselected other phases.

The process is exemplified by the illustration in Figure 4, which
was obtained by processing the same image of the granulite
studied by Kretz (1969) and shown in Plate 1 at the top. An
algorithmic approach proposed by Fabbri, Kasvand, and Masounave
(1982, 1983) allows: (1) the extraction of the binary image of
amphibole and scapolite profiles adjacent to pyroxene profiles,
and that of the pyroxene profiles adjacent only to both amphibole
and scapolite, and (2) separate binary images of the three contacts
amphibole-pyroxene, pyroxene-scapolite, and scapolite-amphibole
from the images extracted in (1). Figure 5 shows the image of the
three phases considered before the extraction mentioned in (1),
(a), and after (b). Such images and contacts are not only
qualitatively, but also quantitatively unavailable in general to the
petrologist who cannot focus attention on more than a few grains
at a time in the field of the microscope.

A number of other chains can be extracted for complete
characterization of the texture. In the instance shown in Figure 5,
the selected profiles may represent reactions (replacement
patterns between pyroxene and scapolite to produce amphibole)
delineating a structure that can be a relict of original layering
retained by the granulite after metamorphism has taken place.
The problem of defining which aspects of grain distribution
prevail in defining the Markovian property which occurs in his
rock by Whitten and others (1975) can be approached in
morphological and statistical terms whereas the physical
appearance of particular Markovian chains can be studied by
extracting information about all grains contributing to it.

Application #3: Frequency of Contacts and of Their Lengths in
Crystalline Aggregates of Granites

In this application two images, shown in Plate 1 at center and
bottom, digitized from thin sections of different granitic rocks
were processed to obtain quantitative information about the
structure of the grain arrangements and to provide detailed data
upon which further statistical analysis could be based. For

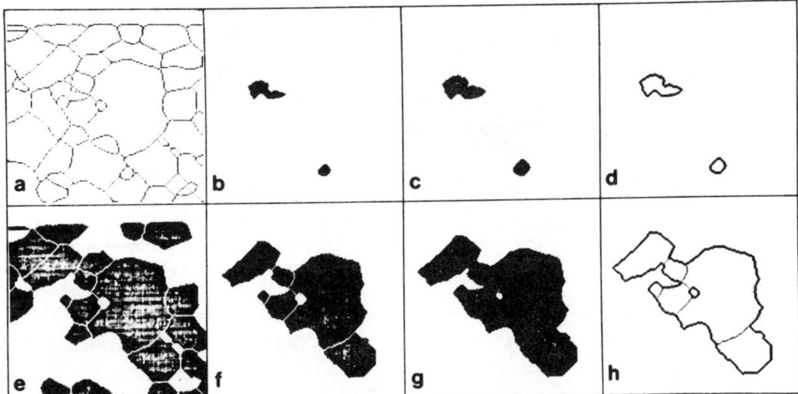

Figure 4. Extraction of pyroxene grain profiles adjacent to two
           amphiboles: (a) 150 x 150 pixel portion of thinned
           boundary binary image of granulite; (b) two amphibole
           grain profiles extracted from image in (a); (c) image of
           dilatated amphiboles; (d) image of white pixels that
           turned black when dilatating image in (b); (e) image of
           pyroxene grain profiles extracted from image in (a); (f)
           image of only pyroxene profiles which are adjacent to
           amphiboles extracted by "shell" image in (d); (g)
           dilatation of image in (f). These pixels will be pointers
           to both amphibole and scapolite profiles adjacent to
           grains of pyroxene in image in (f).

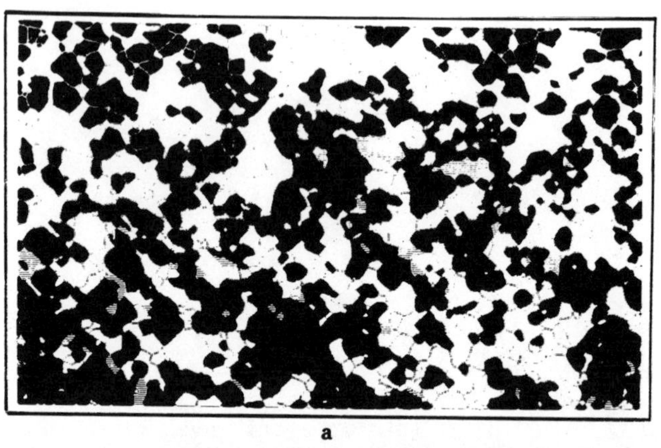

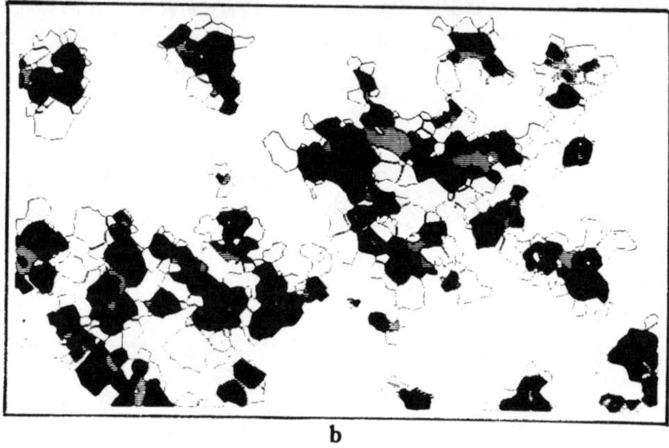

Figure 5. Extraction of pattern of adjacency. Binary images of
hornblende (horizontal ruling), pyroxene (black), and
scapolite (white) before (s) and after (b) extraction of
pyroxene adjacent to hornblende, hornblende adjacent
to pyroxene, and scapolite adjacent to extracted
pyroxene.

example, the randomness of the grain texture and the possible
relationship between permeability and texture were explored.
This was done by Chung and others (1984), after a detailed study
by Kamineni and Katsube (1982), in order to evaluate a host rock
for suitability of radioactive waste disposal by studying the
relationship between permeability and texture of granitic plutons
with highly contrasting permeabilities. The samples, represented
by the phase-labeled images shown in Plate 1, center and bottom,
have permeabilities of 20.1 and 0.3 microdarcies under
unconfined pressure, respectively (4.2 and 0.1 microdarcies
under unconfined in situ pressure). In other words, the rock
corresponding to the image at the center of Plate 1, is
approximately 40 times more permeable than the one
corresponding to the image at the bottom of the plate.

To obtain information on the frequency of contacts between all
pairs of phases present (potash feldspar, plagioclase feldspar,
quartz, biotite, and other minerals), and on contact lengths,
segment-and-junction labeled images were computed, similar to
the one shown in Figure 2. Table 2 shows a portion of a larger list
of data organized according to segment-label numbers. From such
a list, obtained by using also a component-labeled image, and a
phase-labeled image, various statistics can be computed for
obtaining frequency distributions of the segments and their
lengths.

Differences between frequencies of quartz-quartz contacts and in
contact lengths between potash feldspar and potash feldspar seem
to be related to permeability differences. Higher numbers of
quartz-quartz contact contribute to higher permeability due to the
higher number of microfractures in quartz.

The study performed by Chung and others (1984) developed
statistical tests for comparing the pairwise contact frequencies
and lengths observed with the respective values to be expected in
random situations. Such randomness tests for neighborhood
relationships between crystal profiles are feasible.

By using the images of the types shown in Figures 1 and 2
(component-labeled image, phase-labeled image, and
segment-and-junction-labeled image) additional tabulations can be
computed according to component labels, phase labels, and

Table 2. Simplified portion of computer printout of tabulated data
organized according to segment-label number. Column
labels are: (1) table address, (2) marker, (3) segment-
label number, (4) number of pixels on segment, (5)
adjacent component label on one side of segment, (6)
adjacent component label on other side of segment, (7)
and (8) two junction labels at either end of segment, (9)
and (10) adjacent phase labels on either side of segment
corresponding to (5) and (6), and (11) "true" segment
length computed as accurately as possible in digital
image. In table, segments were not labeled in
continuous sequence 2, 3, 4, ...

(1)	(2)	(3)	(4)	(5)	(6)	(7)	(8)	(9)	(10)	(11)
2	1	2	58	3	7	-2	-7	5	2	67.11227
12	1	12	67	7	6	-3	-8	2	3	77.76904
13	1	13	40	7	5	-2	-3	2	3	43.10651
20	1	20	7	5	3	-2	-4	3	5	7.62132
25	1	25	11	6	5	-3	-5	3	3	12.03553
27	1	27	16	7	3	-4	-6	2	5	19.31367
29	1	29	7	3	9	-6	-7	5	4	8.44974
31	1	31	7	6	10	-5	-9	3	6	9.07106
32	1	32	18	10	5	-5	-12	6	3	20.27814
33	1	33	41	7	9	-6	-7	2	4	46.38464
35	1	35	54	7	14	-8	-10	2	4	63.94075
47	1	47	23	14	6	-8	-15	4	3	29.21313
48	1	48	21	6	12	-9	-17	3	5	25.76340
49	1	49	11	12	10	-9	-11	5	6	12.44974
58	1	58	9	7	15	-10	-13	2	4	10.24264
59	1	59	20	14	15	-10	-23	4	4	27.24866
--	-	--	--	--	--	---	---	-	-	--------
--	-	--	--	--	--	---	---	-	-	--------
--	-	--	--	--	--	---	---	-	-	--------

junction labels to address directly the number of pixels in each
crystal profile (component) and in each phase, and to have
information available on which boundary segments are
surrounding each crystal, etc. The tabulated information may be
sorted according various criteria, in order to collect the desired
results.

CONCLUDING REMARKS

This review of crystallization models and applications to grain
adjacency relationships has shown some ways of capturing
microscopic data in digital form and of computing transformations
by image processing. Several ideas, which are not exhaustive,
have been suggested for quantitative characterization.

New developments in simulation of textures by image processing
that can be of interest in petrology and quantitative microscopy
have been suggested by Ahuja, Dubitski, and Rosenfeld (1980).
They experimentally analyzed digital and statistical models of
mosaics representing tessellation in space. Cross and Jain (1983)
explored the use of Markov random fields in texture models.

The study of crystalline fabrics, either for grain-growth pattern
understanding or for associating physical properties to textural
characterization is beginning to expand. Its feasibility will
increase with the development of methods in image processing
and in geometric probability.

REFERENCES

Ahuja, N., Dubitzki, T., and Rosenfeld, A., 1980, Some
            experiments with mosaic models for images: I.E.E.E.
            Trans. Systems, Man, and Cybernetics, v. SCM-10, p.
            744-749.

Chung, C.F., Fabbri, A.G., Kasvand, T., and Otsu, N., 1984, Grain
            contiguity and inference on permeability of granites:
            Proc. ICAM84, Intern. Congress on Applied
            Mineralogy in the Mineral Industry, Los Angeles,
            California, p. 223-244.

Cross, G.R., and Jain, A.K., 1983, Markov random field texture
          models: I.E.E.E. Trans. Pattern Analysis and Machine
          Intelligence, v. PAMI-5, p. 25-39.

Ehrlich, R., Vogel, T.A., Weinberg, B., Kamilli, D.C., Byerly, G., and
          Richter, H., 1972, Textural variations in petrogenetic
          analyses: Geol. Soc. America Bull., v. 83, no. 3,
          p. 665-676.

Fabbri, A.G., 1984, Image processing of geological data: Van
          Nostrand Reinhold Co., New York, 244p.

Fabbri, A.G., and Kasvand, T., 1981, Image processing for the
          detection of two-dimensional Markovian properties as
          functions of distances from crystal profiles: Proc. 3rd
          European Symp. for Stereology, Ljubljana, Yugoslavia,
          p. 153-163.

Fabbri, A.G., Kasvand, T., and Masounave, J., 1983, Adjacency
          relationships in aggregates of crystal profiles: Proc.
          NATO Adv. Study Inst. on Pictorial Data Analysis,
          Bonas, France, in Haralick, R.M., ed., NATO ASI
          Series, v. 4, Springer-Verlag, New York, p. 449-468.

Flynn, D., 1969, Grain contacts in crystalline rocks: Lithos, v. 2,
          no. 4, p. 361-370.

Jen, L.S., 1975, Spatial distribution of crystals and phase
          equilibria in charnockitic granulites from the
          Adirondack Mountains, New York: unpubl. doctoral
          dissertation, Univ. Ottawa, 248p.

Kamineni, D.C., and Katsube, T.J., 1982, Hydraulic permeability
          differences between granites from the Lac du Bonnet
          (Manitoba) and Eye-Dashwa (Ontario) plutons, related
          to textural effects: Current Research, Part a, Geol.
          Survey Canada Paper 82-1A, p. 393-401.

Kretz, R., 1966, Grain size distribution for certain metamorphic
          minerals in relation to nucleation and growth: Jour.
          Geology, v. 74, no. 2, p. 147-173.

Kretz, R., 1969, On the spatial distribution of crystals in rocks:
          Lithos, v. 2, no. 1, p. 39-65.

Miller, P.R., Reid, A.F., and Zuiderwyk, M.A., 1982, QEM*SEM image analysis in the determination of modal essays, mineral association and mineral liberation: Proc. XIV Intern. Mineral Processing Congr., Toronto, Oct. 17-23, 1982, Paper VIII-3, p. 1-20.

Reid, A.F., Gottlieb, P., MacDonald, K.J., and Miller, P.R., 1984, QEM*SEM image analysis of ore minerals: volume fraction, liberation, and observational variances: Proc. ICAM84, Intern. Congr. on Applied Mineralogy in the Mineral Industry, Los Angeles, California, Feb. 22-25, 1984, p. 191-204.

Vistelius, A.B., 1972, Ideal granite and its properties: I. The stochastic model: Jour. Math. Geology, v. 4, no. 1, p. 89-102.

Vistelius, A.B., Agterberg, F.P., Divi, S.R., and Hogarth, D.D., 1983, A stochastic model for the crystallization and textural analysis of a fine-grained granitic stock near Meech Lake, Gatineau Park, Quebec: Geol. Survey Canada Paper 81-21, 62p.

Whitten, E.H.T., Dacey, M.F., and Thompson, K., 1975, Markovian grain relationships of a Grenville granulite: Am. Jour. Sci., v. 275, no. 10, p. 1164-1182.

APPLICATION OF FUZZY SETS TO THE SUBDIVISION OF GEOLOGICAL UNITS

Marek Kacewicz

Warsaw University

ABSTRACT

An automatic subdivision of the lithologic series into units is difficult because of the complexity of available geological information. Classical algorithms usually are based on the assumption that all values of the considered features are given in the numerical form. It is convenient to simplify the analytical process, but it is connected with associate error. Features described in other forms seem to bring us more information about objects. For example, the subject concerns linguistic form - words or sentences from natural or artificial languages, or new forms of describing of data - that is fuzzy sets. We assume that the geological environment is represented by a given set of features, and three forms of describing of data are allowed: numerical, linguistic, and a fuzzy one. Every sample is characterized by a vector. We introduce distances for the considered types of characters and unify their values by transformation to the unit interval. It enables us to determine a simple formula to calculate distances between samples and as a result an algorithm for subdivision of the investigated lithologic series. The algorithm was programmed in FORTRAN IV language for engineering geology problems.

INTRODUCTION

An automatic subdivision of the investigated lithologic series into units is a difficult problem because of the complexity of information which we obtain from geological studies. Usually we consider many different features characteristic of a given geological environment, among which we can select more-or-less important ones from the point of view of a given application. Classical algorithms usually are based on the assumption that all values of considered features are given in the numerical form. It is according to the tacit assumption existing in the science that we cannot acknowledge the process, before it is described in quantitative terms. It is convenient to simplify the analytical process but is connected with the associated error. In documentations the main part of values of these features are assumed subjectively by the researcher and described in such a manner hinders automation of the analytical process. For example it concerns such situations where quantitative and qualitative characters exist in parallel. It is not easy to classify objects described in such a manner.

Methods of numerical taxonomy (Sneath and Sokal, 1973) are useful but do not consider to the proper degree the fact that the information is not precise. In such situations statistical methods are used. Thus the process of subdivision of the lithologic series must be started in the descriptive phase of the project. Proper and good description depends on using such language that allows us to take into consideration the impreciseness and subjective point of view of the researcher. It can allow us to obtain more information than we can from classical methods, where we must give explicit answers in situations when we are not sure. Such language, enabling the description and proper processing of subjectively created, imprecise data, has been researched for years. Zadeh's (1965) work gave us a new useful tool - the theory of fuzzy sets. The theory is useful in such different scientific fields as psychology (Oden, 1977), informatics (Tanaka and Mizumoto, 1975), and decision processes (Fung and Fu, 1975). In geology, it has been applied to define kriging (Rao and Prasad, 1982) and to solve the linear unmixing problem (Full, Ehrlich, and Bezdek, 1982). The theory has been created to verify tacitly the accepted assumption that uncertainty is equivalent to randomness. Classical investigations use probabilistic methods to process imprecise data. Zadeh, the theory creator, claims that uncertainty is not due always to randomness but may originate from other sources. He introduces a certain special class of sets where there is no distinct limit between belonging and not belonging of objects to them. The set of such property is termed a fuzzy set.

Usually we identify fuzzy set with the following function:

$$\mu : A \rightarrow [\, 0,\, 1] \qquad\qquad (1)$$

(where A is a subset of X - the space of certain elements, for example, investigated notions or numbers), which we term a membership function or function defining a fuzzy set (marked A). If $X \supset A$ is finite, we describe the fuzzy set as:

$$A = \left\{ \frac{\mu(a_1)}{a_1},\ \frac{\mu(a_2)}{a_2},\ \ldots,\ \frac{\mu(a_n)}{a_n} \right\} \qquad (2)$$

where: a_i - is an element of A,

$\mu(a_i)$ - value of the membership function on a_i.

For two fuzzy sets A, B are defined fuzzy operations and relations (Bezdek, 1981): containment, equality, complement, intersection, and union. The theory allows us to describe imprecise data or numbers. For example "about 5" we can define as:

$$\text{" about 5 "} = \left\{ \frac{0.1}{3},\ \frac{0.5}{4},\ \frac{1.0}{5},\ \frac{0.5}{6},\ \frac{0.1}{7} \right\}$$

We have opportunity to perform arithmetic operations on such numbers. Development of fuzzy classification methods takes place in parallel to the development of fuzzy-set theory. A pioneering application of fuzzy sets to cluster analysis was made by Ruspini; the work of Dunn and Bezdek with their fundamental algorithm ISODATA was published in 1973. Since then, the theory of fuzzy clustering has been developing rapidly indicating its usefulness in the context of imprecisely defined data. Many fuzzy clustering algorithms are given in Bezdek (1981), and for geological problems such methods were programmed by Bugayets, Zhaumitov, Lobova, and Maximenko (1979).

The main aim of the paper is to present the theoretical basis of a computer clustering program designed to subdivide lithologic series into units. The method differs from classical methods of fuzzy clustering, and methods are introduced for comparing samples assuming that fuzzy sets were used in the phase of description of features.

UNCERTAINTY IN GEOLOGICAL INVESTIGATIONS

There are several reasons for uncertainty and it is important to distinguish them. Bezdek (1981) identified uncertainty due to:

>   (i) inaccurate measurements,
>   (ii) random occurrences, and
> (iii) vague descriptions.

Undoubtedly, there are other philosophically distinguishable categories, but these three provide for an adequate description for the manifestation of uncertainty in deterministic, probabilistic, and fuzzy models.

Probabilistic and fuzzy methods are useful in geology. The theory of fuzzy sets may help geologists describe imprecise information, for example, from boreholes. From the point of view of engineering geology, sample features that have a fuzzy character include soil type, moisture, and state of the soil. An experiment executed at Warsaw University is given as an example.

It is known that macroscopic investigations of soils are imprecise. Usually the geologist is obligated to give an explicit answer although this may be impossible. Thus a test was constructed that allowed the geologist to describe his inaccuracy. Ten soil samples were selected in such a way that they could be difficult to interpret, and the following table was prepared:

Z	P_o	P_r	P_s	P_d	P_π	P_g	π_p	π	G_π	G	G_p	G_{pz}	G_z	$G_{\pi z}$	I_π	I_p	I

where in the first row there are symbols of soils according to the Polish norm. Then ten geologists were asked to answer the question to what degree do the symbols satisfy samples. Their answers depended on putting numbers from the unit interval to every position of the second row of the table. The result of this operation was the construction of a fuzzy set. For example, the following sets were obtained:

$$\bar{A}_1 = \left\{ \frac{0.2}{G_\pi}, \frac{0.2}{G_{p z}}, \frac{0.3}{G_z}, \frac{0.8}{G_{\pi z}} \right\} ,$$

$$\bar{A}_2 = \left\{ \frac{0.2}{\pi_p}, \frac{0.8}{\pi}, \frac{0.4}{G_\pi} \right\} .$$

Operations of intersection, union, and measures of fuzziness described by Yager (1979), allowed a conclusion about common thinking of the geologists, and all the possibilities and degree of uncertainty of the given answers. On this basis conclusions were drawn about the type of the samples. For example, from the union set for the forth sample:

$$\bar{A}_4 = \bigcup_{i=1}^{10} A_{4i} = \left\{ \frac{1}{P_g}, \frac{0.9}{G_p}, \frac{0.5}{G_\pi}, \frac{0.7}{G}, \frac{0.4}{\pi_p} \right\}$$

it may be concluded that values of the membership function 1, 0.9, 0.7 indicate the limits between soil types denoted by P_g, G_p, and G (in the Feret triangle). The limitation of space in this paper precludes presentation of the full details of this test, but they will be published later. The result of this test shows the fuzzy character determination of the type of soil and as a consequence some geotechnical parameters. Analogously, the fuzzy character of other mentioned features also can be shown, which emphasized that the consideration of fuzzy form of description of data has a practical sense.

SUBDIVISION OF THE INVESTIGATED LITHOLOGIC
SERIES INTO UNITS

Let us assume that in a given geological environment n characteristics are distinguished features. We mark U_1, U_2, ..., U_n sets containing all states of these features. Every U_i may be a set of numbers, linguistic term, or fuzzy sets. Let us consider the Cartesian product:

$$U = U_1 \times U_2 \times ... \times U_n \qquad (3)$$

We should consider subdivisions of the investigated lithologic series as subsets of U. Let us assume that we have m samples A_1, ..., A_m from a borehole characterized by vectors

$\vec{A}_1, \vec{A}_2, ..., \vec{A}_m \in U$ where

$$\vec{A}_1 = (a_{11}, a_{12}, ..., a_{1n})$$
$$\vec{A}_2 = (a_{21}, a_{22}, ..., a_{2n})$$
$$..$$
$$\vec{A}_m = (a_{m1}, a_{m2}, ..., a_{mn}) \qquad (4)$$

and a_{ij} is a state of j-th feature for i-th sample. Our task is to determine the distance between $\vec{A}_i$ and $\vec{A}_j$ for i, j = 1, 2, ..., m. Because of possible differences among U_i, U_j, for i, j = 1, 2, ..., n, i $\neq$ j, it is difficult to define identical measures of distances for all i. For certain i those distances are among real numbers, for others between linguistic terms or fuzzy sets. We distinguish in our program two methods of solving of this problem. The first method depends on introducing measures of distances separately for each coordinate and after unifying their values calculating distances between samples. The second method depends on description of all features in the form of fuzzy sets, and calculation distances between samples using unifying method.

Method 1

Calculation of distances:

Case 1: Let us assume that values of an i-th feature are real numbers, thus $a_{ij} \in R$. Measure of distance is defined as usual:

$$d_i = |\, a_{li} - a_{ji}\, | \qquad \text{for } j, l = 1, 2, \ldots, m \qquad (5)$$

Case 2: Let B_1 and B_2 are fuzzy sets. Then B_1, B_2 obviously may not contain the same elements. From the fuzzy-set theory we know that those values of B_1 which not belong to B_2, may be attached to it with value of membership function equal 0. In the same manner we can enlarge B_1. Then we can calculate distance between two fuzzy sets using one of the following formulae:

$$d(\bar{B}_1, \bar{B}_2) = \frac{\left(\sum_{i=1}^{N} \left| \mu^i_{\bar{B}_1} - \mu^i_{\bar{B}_2} \right|^p \right)^{1/p}}{(N)^{1/p}}, \quad p=1,2,\ldots \quad (6)$$

$$d(\bar{B}_1, \bar{B}_2) = \sup_{i=1,\ldots,m} \left| \mu^i_{\bar{B}_1} - \mu^i_{\bar{B}_2} \right|, \qquad (7)$$

where N - is a number of the belonging to both sets elements after the operation of the enlargement.

$\mu^i_{\overline{B}_1}$ - value of the membership function for i-th $\overline{B}_1$ element

$\mu^i_{\overline{B}_2}$ - value of the membership function for i-th $\overline{B}_2$ element.

<u>Case 3</u>: Let us assume that U_k is a set of terms of a linguistic type, and we have a number scale from 0 to M step 1. Values of distances are evaluated subjectively by the geologist. For example we can say that:

$$d_k(\text{VERY LOW, VERY HIGH}) = M,$$
$$d_k(\text{MH, SM}) = 10. \qquad \text{etc.}$$

After calculating distances for all coordinates, we unify their values by transformation to the unit interval. It let us evaluate distances between each pair of samples. If $A_i = (a_{i1}, a_{i2}, ..., a_{in})$, $A_j = (a_{j1}, a_{j2}, ..., a_{jn})$ represent samples A_i and A_j respectively, and $d_1(a_{i1}, a_{j1})$, $d_2(a_{i2}, a_{j2})$, ..., $d_n(a_{in}, a_{jn}) \in <0,1>$ are unified values then we can calculate distance between these two samples according to formulae:

$$D(A_i, A_j) = \frac{\left(\sum_{k=1}^{n} \overline{d}_k^{\,p} \right)^{1/p}}{(n)^{1/p}}, \quad p = 1, 2, ... \qquad (8)$$

or

$$D(A_i, A_j) = \max_{k=1,2,...,n} \overline{d}_k \qquad (9)$$

Method 2

We describe all features in the form of fuzzy sets. For numbers it
is simple because for every $x \in R$ we may write in the form

$$\left\{ \frac{1.0}{x} \right\} .$$

In the situation of linguistic terms, the geologist subjectively
creates corresponding fuzzy sets. For example:

$$\text{"SMALL"} = \left\{ \frac{1.0}{0}, \frac{0.8}{0.25}, \frac{0.5}{0.3}, \frac{0.2}{0.6} \right\} ,$$

$$\text{"WET"} = \left\{ \frac{0.1}{11}, \frac{0.3}{15}, \frac{0.5}{20}, \frac{0.8}{25}, \frac{1.0}{30} \right\} , \quad \text{etc.}$$

Distances between fuzzy sets and samples are calculated according
to (6) or (7) and (8) or (9), respectively.

PROGRAM

In the program, Method 1 and Method 2 may be selected
alternatively. As an example, a sequence of a borehole with
marked distances between samples is presented in Figure 1,
where D_{ij} denote $D(A_i, A_j)$.

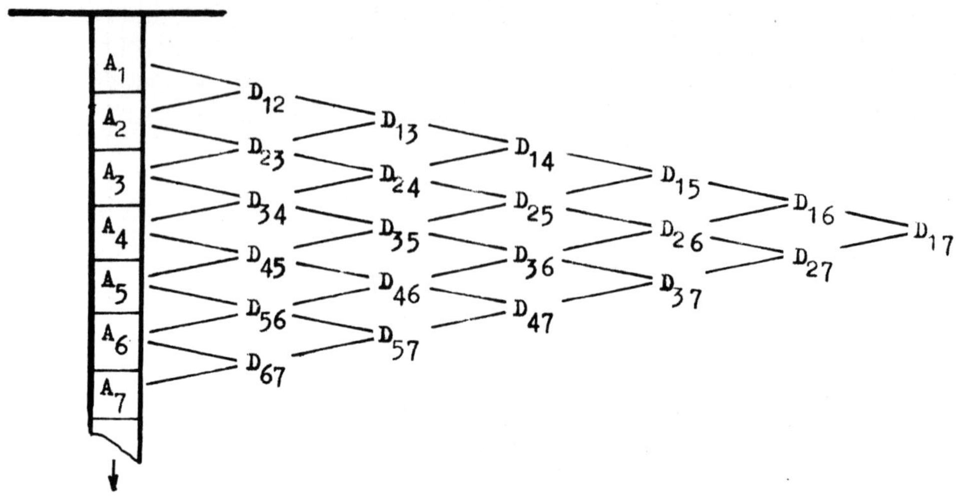

Figure 1. An example sequence of borehole and distances
          between samples

In such a manner we obtain the similarity matrix:

$$
\begin{array}{c|cccc}
 & A_1 & A_2 & \cdots & A_m \\
\hline
A_1 & D_{11} & D_{12} & \cdots & D_{1m} \\
A_2 & D_{21} & D_{22} & \cdots & D_{2m} \\
\vdots & \vdots & \vdots & \cdots & \vdots \\
A_m & D_{m1} & D_{m2} & \cdots & D_{mm}
\end{array}
$$

where $D_{ij} = D_{ji}$.

Notice that the first diagonal above the main diagonal gives certain
information concerning continuity of divisions according to a
selected set of features. The second and other diagonals give
information about the speed of variability. Now we shall introduce
the definition of a unit.

DEFINITION 1

The set of samples $W = \{A_i, A_{i+1}, ..., A_k\}$ is a unit if for arbitrary selected $\varepsilon \in <0,1>$, holds $D_{jl} \leq \varepsilon$ for $j,l = i, i+1, ..., k$.

The block diagram of the algorithm is presented in Figure 2.

AN EXAMPLE

	A_1	A_2	A_3	A_4	A_5	A_6	A_7
A_1	0	0.2	0.2	0.4	0.6	0.2	0.1
A_2	0.2	0	0.3	0.3	0.5	0.7	0.2
A_3	0.2	0.3	0	0.4	0.2	0.4	0.3
A_4	0.4	0.3	0.4	0	0.1	0.2	0.5
A_5	0.6	0.5	0.2	0.1	0	0.2	0.3
A_6	0.2	0.7	0.1	0.2	0.2	0	0.2
A_7	0.1	0.2	0.3	0.5	0.3	0.2	0

Step 1: For selected $\varepsilon = 0.4$ obtained areas are marked in the matrix. We get the following units: $W_1=\{A_1,A_2,A_3,A_4\}$, $W_2=\{A_3,A_4,A_5,A_6\}$, $W_3=\{A_5,A_6,A_7\}$. Notice that A_3 and A_4 are common for W_1 and W_2; A_5 and A_6 are common for W_2 and W_3.

Step 2: For $\varepsilon = 0.3$ we get: $W_1=\{A_1,A_2,A_3\}$, $W_2=\{A_4,A_5,A_6\}$, $W_3=\{A_5,A_6,A_7\}$. The intersection of W_1 and W_2 is empty; A_5 and A_6 are common for W_2 and W_3.

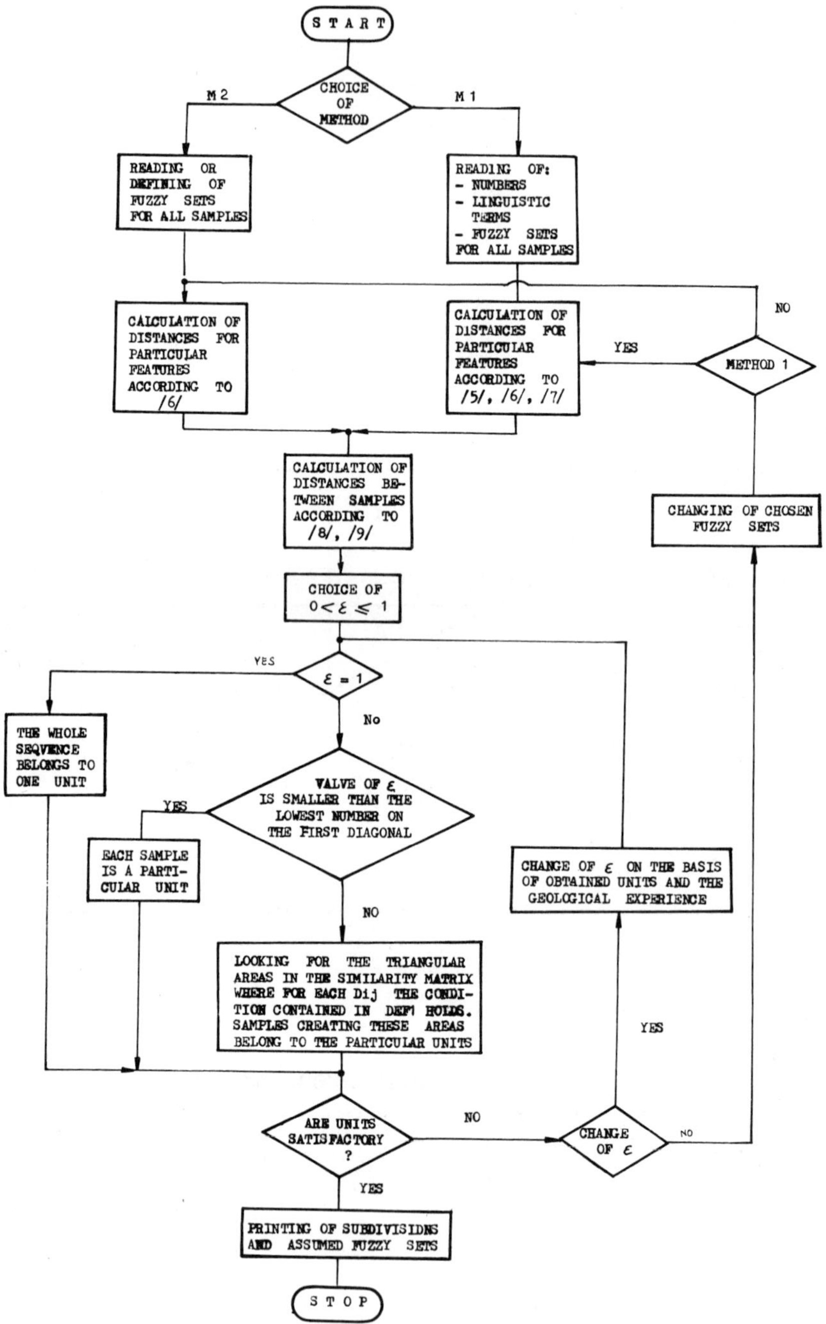

Figure 2. Block diagram of subdividing algorithm.

<u>Step 3</u>: For $\varepsilon = 0.2$ we get: $W_1=\{A_1,A_2\}$, $W_2=\{A_3\}$, $W_3=\{A_4,A_5,A_6\}$. W_3 and W_4 have nonempty intersection.

CONCLUSIONS

The basis of our interactive computer program for engineering-geological problems has been presented. It was used in a study of the Belchatow brown-coal, open-pit mine giving satisfactory results not only from a theoretical point of view, but also for engineering-geological practice.

In summary it has been noted that:

- the theory of fuzzy sets allows us to describe imprecise geological information,

- the results of the test prove that there exist features having fuzzy character in the engineering geology,

- the presented algorithm allows consideration different types of descriptions of features simultaneously,

- interactive character of the program allows modeling subdivisions online, and the

- necessity of defining of fuzzy sets for every sample and distances for linguistic terms makes execution of the program long.

REFERENCES

Bezdek, J.C., 1981, Pattern recognition with fuzzy objective function algorithms: Plenum Press, New York and London, 256 p.

Bugayets, A.N., Zhaumitov, B.Zh., Lobova, O.Y., and Maximienko, L.A., 1979, The software for solving problems of ore deposits: Hornicka Pribram ve Vede a Technice, Pribram, p. 715-721.

Full, W.E., Ehrlich, R.E., and Bezdek, J.C., 1982, FUZZY QMODEL -
    a new approach for linear unmixing: Jour. Math. Geology,
    v. 14, no. 3, p. 259-270.

Fung, L.W., and Fu, K.S., 1975, Axiomatic approach to rational
    decision making in a fuzzy environment, in Fuzzy sets and
    their application to cognitive and decision processes:
    Academic Press, New York, p. 227-256.

Oden, G.C., 1977, Fuzziness in semantic memory: Memory &
    Cognition Jour., v. 5, no. 2, p. 198-204.

Rao, S.V.L.N., and Prasad, J., 1982, Definition of kriging in terms
    of fuzzy logic: Jour. Math. Geology, v. 14, no. 1, p.37-42.

Sneath, P.H.A., and Sokal, R.R., 1973, Numerical taxonomy: W.H.
    Freeman and Co., San Francisco, 572 p.

Tanaka, K., and Mizumoto, M., 1975, Fuzzy programs and their
    execution, in Fuzzy sets and their application to cognitive
    and decision processes: Academic Press, New York,
    p. 41-76.

Yager, R.R., 1979, On the measure of fuzziness and negation. Part
    1: Membership in the unit interval: Rept. RRY 79-016, Sch.
    of Business Administration, New Rochelle, New York, 15 p.

Zadeh, L.A., 1965, Fuzzy sets: Inform. and Control, v. 8, no. 3,
    p. 338-353.

QUALITY OF TIME SCALES - A STATISTICAL APPRAISAL

Frederik P. Agterberg

Geological Survey of Canada

ABSTRACT

A primary concern of geologists is the relative ordering of events
in Earth history. On a regional basis, spatial relationships of
separate or overlapping rock bodies are used for accomplishing
this goal. For correlation over large distances between regions or
when the rate of change of geological processes is being
considered, it is necessary to use the numerical time scale which
is mainly based on radiometric dates of variable. precision.
Existing statistical models for estimating the ages of
chronostratigraphic boundaries and the corresponding error bars
are reviewed. An alternative approach based on the method of
maximum likelihood is presented. Computer simulation
experiments are described to compare estimates obtained by
different methods from small samples of age determinations. A
spline-fitting technique can be used to combine estimates for
successive stage boundaries with one another. The Jurassic time
scale is used to exemplify this statistical approach.

INTRODUCTION

The relative ordering of events in earth history is a primary
concern of geologists. On a regional basis, spatial relationships of
separate or overlapping rock volumes (stratigraphic correlation)
are used for accomplishing this goal. The simplest type of relative

time scale is a sequence of ordered events. From the variable amounts of overlap between rock volumes, or by making assumptions on rates of sedimentation, it may be possible to estimate intervals between events along a relative time axis.

For precise correlation in time over large distances between regions or when the rate of change of geological processes in time is being considered, it is necessary to use the numerical time scale which is based on a combination of chronostratigraphic information with radiometric ages of variable precision. This paper is concerned with methods to estimate the age of stage boundaries in the geologic time scale.

Two new time scales have been published recently (Odin, 1982; and Harland and others, 1982). There is general agreement of the ages along most of these time scales. The largest discrepancies amount to about 10 percent of the ages estimated. Harland and others (1982) estimated 144 Ma and 590 Ma, and Odin (1982) 130 Ma and 530 Ma for the Jurassic-Cretaceous and Precambrian-Cambrian boundaries, respectively. These largest differences are related to the nature of the materials used for dating. Glauconite ages near these boundaries tend to be younger than ages obtained from other materials. Harland (1983) believes that Odin's (1982) ages in these instances are too young because "glaucony" in sediments dated by the K-Ar method may have produced ages which are too low due to escape of Ar.

The statistical methods presented in this paper cannot be used to resolve difficulties related to the nature of the materials used for dating. Neither can they contribute significantly to the problem of selecting decay constants in order to avoid bias in radiometric dating. However, any radiometric method is subject to a measurement error which usually is greater than the uncertainties associated with the relative ordering of events using methods of stratigraphic correlation (for example, biostratigraphic or magnetopolarity methods). The problem of having to estimate the age of stage and chronozone boundaries from relatively imprecise isotope determinations would remain even if all sources of bias related to these methods could be eliminated.

Cox and Dalrymple (1967) have developed a statistical approach for estimating the age of boundaries between polarity chronozones in the Cenozoic (Brunhes, Matuyama, Gauss, and Gilbert Chronozones). A slightly modified version of their method was used in Harland and others (1982) for estimating the ages of boundaries between the stages of the Phanerozoic geologic time scale. In the sequel, references to Harland and others (1982) will be denoted as GTS (Geological Time Scale).

The statistical approach in GTS is as follows. Suppose that t_e represents an assumed trial or "estimator" age for the boundary between two stages. Then measured ages t in the vicinity of this boundary can be classified as t_y (younger) or t_o (older than the assumed stage boundary). Each age determination t_{yi} or t_{oi} has its own standard deviation s_i (i=1, 2, ..., n). Because these standard deviations are relatively large, a number (n') of the age determinations may be inconsistent with respect to the estimator t_e. Only the n' inconsistent ages t_i' with $t_{oi}<t_e$ and $t_{yi}>t_e$ were used for estimation by Cox and Dalrymple (1967). These inconsistent ages may be indicated by letting i go from 1 to n'. This implies that there are (n-n') consistent ages labeled (n'+1) to n.

In GTS, a quantity E^2 with

$$E^2 = \sum_{i=1}^{n'} \frac{(t_i' - t_e)^2}{s_i^2}$$

was plotted against t_e in the chronogram for a specific stage boundary.

The plot of E^2 against t_e usually has a parabolic form. The value of t_e for which E^2 is a minimum was used as the estimated age of the stage boundary. A quantitative estimate of the error of this value also was obtained in GTS from the chronogram by taking this error as one-half the age range for which E^2 did not exceed its minimum value by more than 1.0. For convenience, this method of estimating the error will be referred later in this paper as the "error-range method."

If the available data are sparse, inconsistent ages may be absent and E^2 is equal to zero between the highest value of the t_y's and the lowest value of the t_o's. If there are no inconsistent ages, the midpoint of the age range for which $E^2=0$ may be used so that an estimate always is obtained.

It will be shown by the use of computer simulation experiments later in this paper that the preceding method with the addition given in the last paragraph provides an unbiased estimator of the true age which is nearly as good as the maximum likelihood (ML) estimator to be developed. The standard error of this ML estimator is less than the one obtained by the error-range method. However, the procedure of using only the inconsistent ages for estimation may be questioned. In the absence of bias, the ages with the largest errors are more likely to become inconsistent and, therefore, are weighted more strongly than the more precise age determinations in the method followed in GTS. In general, the consistent ages also provide information that is relevant for estimating the age of chronostratigraphic boundaries. The maximum likelihood approach adopted in this paper makes use of consistent as well as inconsistent ages.

A parabolic chronogram is obtained more readily when the consistent ages are used together with the inconsistent ages as will be seen later. A numerical example of the types of differences in results obtained is as follows. The chronogram for the Norian-Rhaetian boundary (Harland and others, GTS, fig. 3.4h) is based on 6 ages. Values of E^2 were estimated with t_e going from 178 Ma to 242 Ma in steps of 2 Ma. Because, $E^2 = 0$ for t_e equal to 212 and 214, the resulting age of the Norian-Rhaetian boundary would be approximately 213 Ma. The corresponding standard error (GTS, table 3.1 and fig. 5.4) is 9 Ma. The maximum likelihood method of this paper (see later) using the same set of 6 data gives an estimated age of 215.5 Ma with corresponding standard error of 4.2 Ma.

Suppose that the ages of a number of successive stage boundaries have been estimated with their standard deviations. In GTS the following method is used to improve further these estimates. In general, if the error-range method gave a standard error of less than 2.5 Ma for stage boundaries that are younger than 200 Ma, and less than 6 Ma for older stage boundaries, these estimates were declared "tie-points". The hypothesis of equal duration of stages was used to make linear interpolations between tie-points replacing chronogram estimates not sufficiently precise to be declared tie-points. For example, the error-range method gave an error of 9 Ma for the Norian-Rhaetian boundary (see before)

which was too large for it to be made a tie-point. Instead of this,
linear interpolation between nearest tie-points gave an estimated
age of 219 Ma (instead of 213 Ma or 215.5 Ma, see before) for the
Norian-Rhaetian boundary.

In the second part of this paper, smoothing by using cubic spline
functions will be considered as a method by which successive age
estimates for stage boundaries can be improved. An advantage of
this approach is that a smooth curve is fitted to values each of
which is weighted according to the inverse of its variance. The
hypothesis of equal duration of zones instead of stages can be used
for spline-fitting; tie-points can be incorporated as points which
are forced to be situated on the smooth curve.

MAXIMUM-LIKELIHOOD METHOD FOR ESTIMATING THE AGE OF STAGE BOUNDARIES

The statistical model originally proposed by Cox and Dalrymple
(1967) may be formulated as follows. Suppose that a stage with
upper age boundary t_1 and lower boundary t_2 is sampled at
random. This yields a population of ages $t_1 < t < t_2$ with uniform
frequency density function h(t). Suppose that every age
determination is subject to an error which is distributed normally
with unity variance. In general, we have for the frequency density
function f(t) of measurements of which the errors satisfy the
density function ϕ for the normal distribution in standard form:

$$f(t) \; = \; \int_{-\infty}^{\infty} \phi\,(t - x)\, h(x) \; dx$$

Because h(t) is uniform, this becomes

$$f(t) \; = \; \frac{1}{t_2 - t_1} \int_{t_1}^{t_2} \phi\,(t - x)\; dx$$

or,

$$f(t) = \frac{1}{t_2 - t_1} \left[\Phi(t - t_1) - \Phi(t - t_2) \right]$$

where Φ represents the cumulative distribution function of the normal distribution in standard form. For this derivation, the unit of t was set equal to the standard deviation of the errors. Alternatively, the duration of the stage can be kept constant whereas the standard deviation(s) of the measurements is changed.

Suppose that $t_2 - t_1 = 1$, then f(t) becomes

$$f(t) = \Phi\left(\frac{t - t_1}{s}\right) - \Phi\left(\frac{t - t_2}{s}\right)$$

Graphical representations of f(t) for different values of s were given by Cox and Dalrymple (1967, fig. 7, p. 2611). It could be argued that h(x) is not necessarily uniform, then f(t) would be different. However, one would need large samples of age determinations before the selection of a different model for h(x) would be justified.

Suppose now that the true age τ_e of a single stage boundary is to be estimated from n measurements of variable precision on specimens which are known to be either younger or older than the age of this boundary. This problem can be solved if a weighting function f(x) is defined such as those shown in Figure 1. The boundary is assumed to occur at the point where x = 0. If we are interested only in the lower boundary of a stage, $\Phi\{(t - t_1)/s\}$ can be set equal to one yielding the third weighting function $f_3(x)$ of Figure 1. Alternatively, this weighting function can be derived directly: If all possible ages above the stage boundary have an equal chance of being represented, then the probability that their measured age assumes a specific value is proportional to the integral of the Gaussian density function for the errors. Although $f_3(x)$ seems to provide the best possible weighting function under

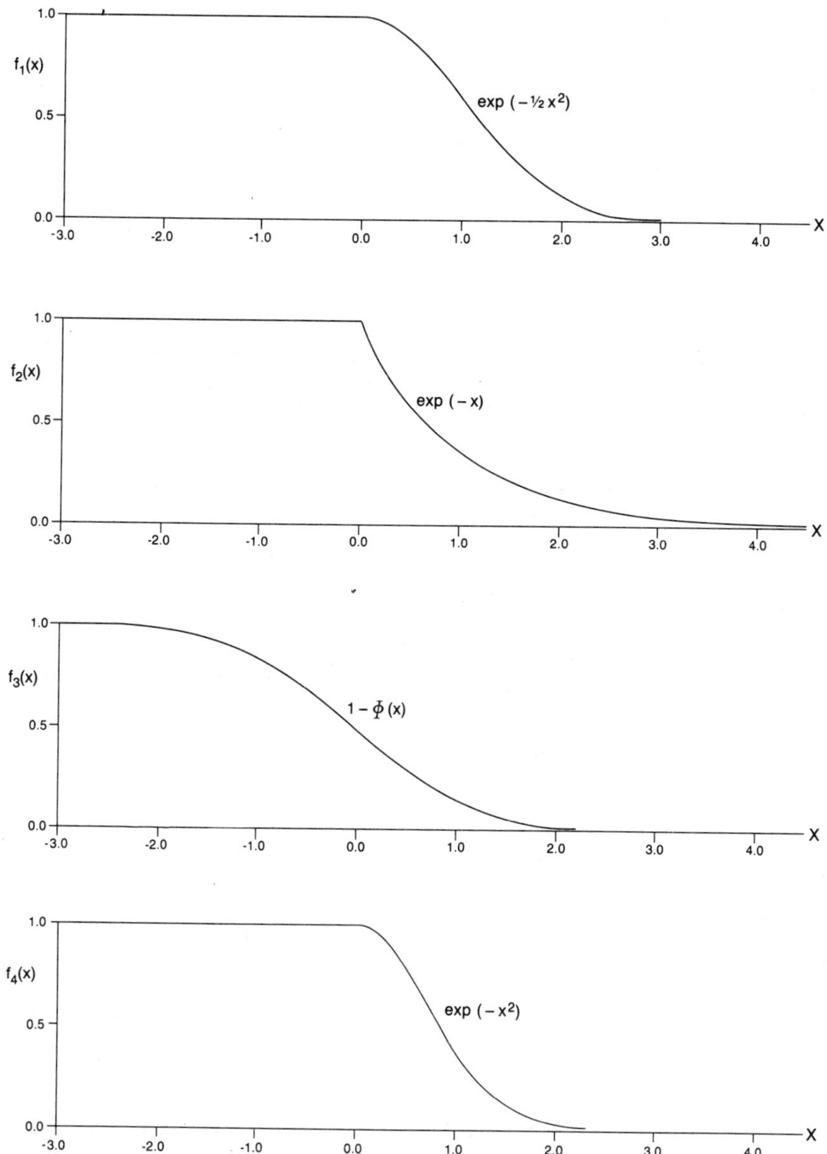

Figure 1   Examples of weighting functions on basis of which likelihood function can be estimated. Function $f_3(x)$ follows from assumption that each age determination is sum of two random variables for (1) uniform distribution of (unknown) true ages, and (2) Gaussian distributions of measurement errors.

the circumstances, the use of several other weighting functions also is considered here. This will allow us to interpret chronograms in GTS as maximum likelihood solutions for the weighting function $f_1(x)$.

In terms of the definitions given in the introduction, any inconsistent age t_y greater than t_e has $x > 0$ whereas consistent ages with $t_y < t_e$ have $x < 0$ in Figure 1. By reversing the positive direction of x for older ages, it follows that any inconsistent age t_o less than t_e also has $x > 0$ whereas consistent ages with $t_o > t_e$ have $x < 0$. It is assumed that standardization of an age t_{yi} or t_{oi} can be achieved by dividing either $(t_{yi} - t_e)$ or $(t_{oi} - t_e)$ by its standard error s_i yielding $x_i = (t_{yi} - t_e)/s_i$ or $x_i = (t_{oi} - t_e)/s_i$.

Suppose that x_i is a realization of a random variable X. The weighting function of $f(x)$ in Figure 1 then can be used to define the probability $P_i = P(X_i = x_i) = f(x_i)\,\Delta x$ that x will lie in a small interval Δx about x_i. The method of maximum likelihood for a sample of n values x_i consists of determining the value of t_e for which the product of the probabilities P_i is a maximum. Because Δx can be set equal to an arbitrarily small constant, this maximum occurs when the likelihood function

$$L = L(x \mid t_e) = \prod_{i=1}^{n} \left\{ f(x_i) \right\}$$

is a maximum.

Kendall and Stuart (1961, p. 43-44) present a mathematical proof that if the first two derivatives of L with respect to t_e exist in an interval of t_e which includes the true value τ_e, then the maximum likelihood estimator is distributed asymptotically normally with mean τ_e at the maximum where

$$\frac{d \log L(x \mid t_e)}{dt_e} = 0$$

and variance

$$s^2\left(\hat{\tau}_e\right) = \frac{-1}{\dfrac{d^2 \log L\left(x \mid t_e\right)}{dt_e^2}}$$

This topic is reviewed in more detail in Appendix 1. It is convenient to use $\log_e L$ instead of L. This is termed the log-likelihood function. If $f(x \mid t_e) = 1$ for $x \le t_e$, the $(n - n')$ consistent measurements do not contribute to the log-likelihood function. Consequently, multiplication of the probabilities or summation of their logs need only be carried out with respect to the n' inconsistent measurements. For the four weighting functions shown in Figure 1, the log-likelihood functions can be written as follows:

For

$$f_1\left(x \mid t_e\right) = e^{-(x-t_e)^2/2} \quad \text{with } x > t_e :$$

$$\log_e L_1\left(x \mid t_e\right) = -(1/2)\sum_{i=1}^{n'} \left(x_i - t_e\right)^2 \qquad (1)$$

For

$$f_2\left(x \mid t_e\right) = e^{-(x-t_e)^2} \quad \text{with } x > t_e :$$

$$\log_e L_2\left(x \mid t_e\right) = \sum_{i=1}^{n''} \left(x_i - t_e\right) \qquad (2)$$

For

$$f_3\left(x \mid t_e\right) = 1 - \Phi(x - t_e) :$$

$$\log_e L_3\left(x \mid t_e\right) = \sum_{i=1}^{n} \log_e \left\{ 1 - \Phi(x_i - t_e) \right\} \qquad (3)$$

For

$$f_4(x|t_e) = e^{-(x-t_e)^2} \quad \text{with } x > t_e :$$

$$\log_e L_4(x|t_e) = -\sum_{i=1}^{n'} (x_i - t_e)^2 \tag{4}$$

From the definitions given in the introduction it can be seen that

$$E^2 = -2 \log_e L_1(x|t_e) = -\log_e L_4(x|t_e)$$

The quantity E^2 is plotted in the vertical direction of Figure 2 for the Caerfai-St. David's boundary example taken from GTS (fig. 3.7i). The data on which this chronogram is based are shown along the top. Values of E^2 were calculated at intervals of 4 Ma and a parabola was fitted to the resulting values by using the method of least squares. If the log-likelihood is parabolic, with E^2 satisfying

$$E^2 = a + bt_e + ct_e^2$$

it follows that the maximum likelihood estimator is distributed normally with mean $\hat{t}_e = -b/2c$ and variance $s^2(\hat{t}_e) = -1/2c$. It will be shown in the next paragraph that graphically $s(\hat{t}_e)$ might be determined by taking one-fourth of the width of the parabola at the point where E^2 exceeds its minimum value by 2 (see Fig. 2). It is noted, however, that the latter result applies to L_4 and not to L_1. If $E^2/2$ were plotted in the vertical direction, the log-likelihood function $\log_e L_1(x|t_e)$ would be obtained with

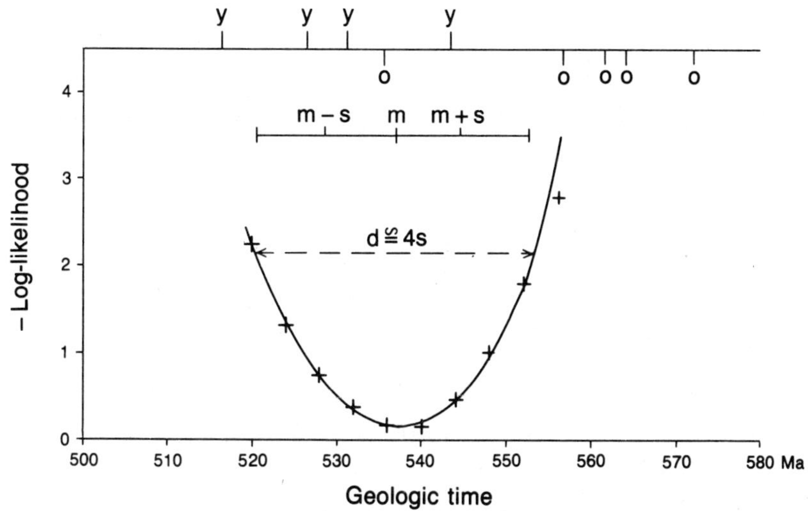

Figure 2. Chronogram for Caerfai-St. David's boundary example
and parabola fitted by method of least squares.
E^2 = - log-likelihood is plotted in vertical direction.
Dates belonging to stages which are older and younger
that boundary are indicated by o and y, respectively.
Standard deviation follows from d representing width
of parabola for E^2 equal to its minimum value
augmented by 2.

variance equal to $2 s^2 (\hat{\tau}_e)$. Consequently, the standard deviation that results from using L_1, is $\sqrt{2}$ times as large as the one obtained from L_4. Harland and others (1982) defined the error of their estimate by taking one-half the age range for which E^2 does not exceed its minimum value by more than 1.0 (error-range method, see Introduction). Because, similar to L_1, this yields a standard deviation that is $\sqrt{2}$ times as large as the one resulting from L_4, it proves that the GTS method is equivalent to a maximum likelihood method with weighting function $f_1(x)$.

A simple proof of the validity of the modified error-range method illustrated in Figure 2 is as follows. According to the theory of mathematical statistics (cf., Appendix 1), the likelihood function is normal asymptotically:

$$e^y = \frac{1}{\sqrt{2\pi}\ \sigma} \exp\left(-t^2/2\sigma^2\right)$$

In this expression $e^y = L(x \mid t_e)$ and $t = t_e - \tau$; σ represents the standard deviation of this normal curve centered about $\tau = 0$. Taking the logarithm at both sides gives the parabola:

$$y = \max - t^2/\ 2\sigma^2$$

where max represents the maximum value of the log-likelihood function. Setting $y = \max-2$ gives $t = 2\sigma$. This indicates that the width of the parabola at 2 units of y below its maximum value is equal to 4σ. The parabola shown in Figure 2 (and subsequent illustrations) is assumed to provide an approximation of the true log-likelihood function. The standard deviation obtained from the fitted curve is written as s. Alternatively, the value of $\max = - \log_e\sigma - (1/2) \log_e 2\pi$ could be used to obtain s as an estimate of σ. In Figure 2, the y-axis has been inverted so that $- y = E^2$ points upwards in order to facilitate comparison with the chronograms in GTS.

Figures 3 and 4 show estimates of $\hat{\tau}_e$ based on $L_2(x \mid t_e)$ and $L_3(x \mid t_e)$, respectively. The purpose of Figure 3 is to illustrate

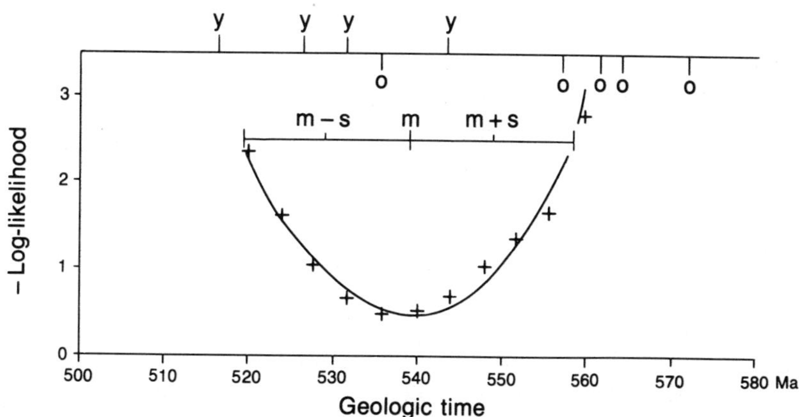

Figure 3. Caerfai-St. David's boundary example assuming that L2 can be used.

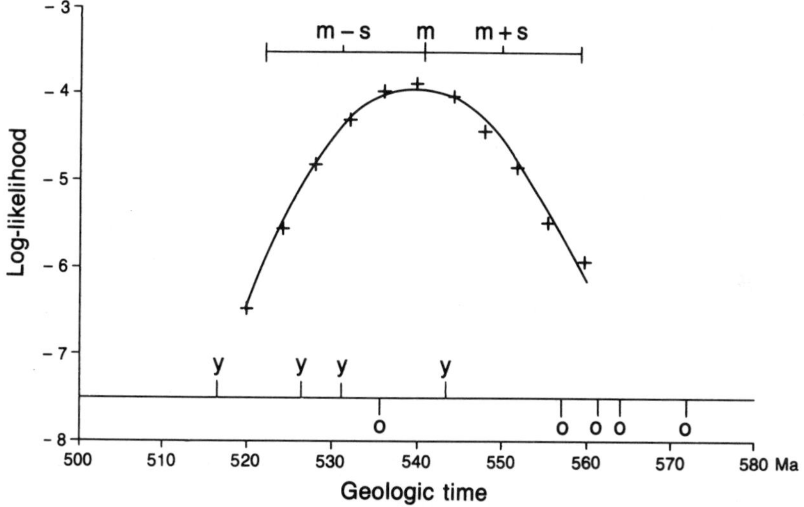

Figure 4. Caerfai-St. David's boundary example. Age (m) estimated by maximum likelihood method using L3. Standard deviation (s) and width of 95 percent confidence interval are approximated closely by results shown in Figure 2.

that a rather large change in the selection of weighting function (f_2 instead of f_4) does not give necessarily different estimates of the mean and standard deviation. The estimated ages of the Caerfai-St. David's boundary and their standard deviations obtained for L_4, L_2, and L_3 (see Fig. 2 to 4) are similar. Although the maximum likelihood estimates of the mean based on L_1 and L_4 are identical, the variance for L_4 is closer to the variance for L_3. It therefore may be concluded that, for approximation purposes, L_4 should be used instead of L_1. This conclusion will be corroborated by a more detailed comparison of the weighting functions for L_1, L_3, and L_4 at the end of this section, and by computer simulation experiments to be described in the next section. However, L_4 does not provide a good approximation of L_3 when inconsistent ages are missing. The Norian-Rhaetian boundary (GTS, fig. 3.4h) provides an example of this. Figures 5 and 6 show results for L_3 and L_4, respectively. Although the age estimated in Figure 6 probably is almost as good, the corresponding standard deviation is larger than that in Figure 5 because the approximation of E^2 by a parabola is not good.

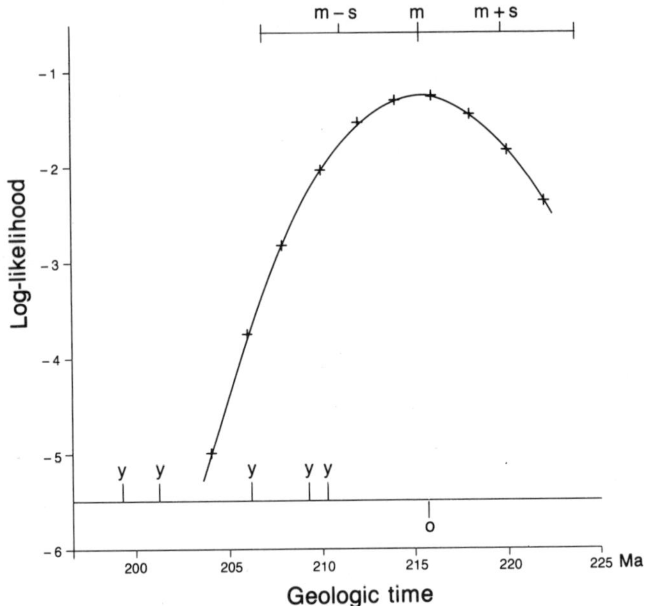

Figure 5. Norian-Rhaetian boundary example. Log-likelihood function using L_3 has converged to parabola although only 6 consistent dates were available.

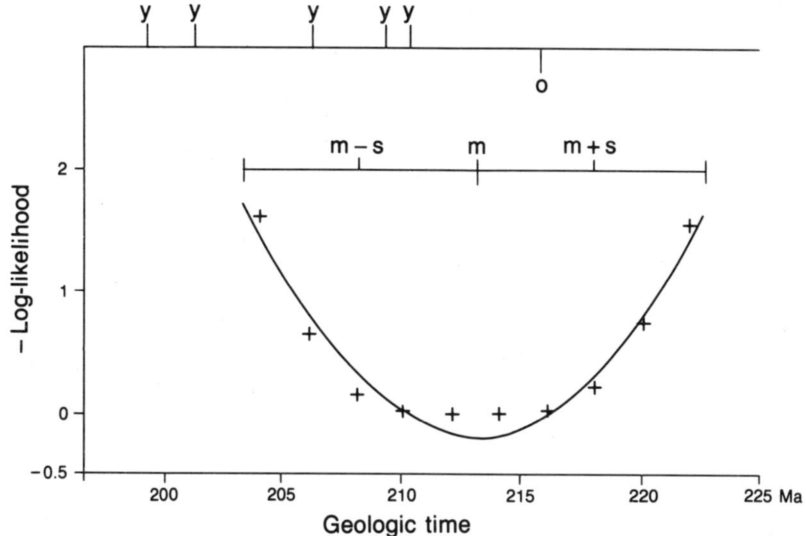

Figure 6.   Norian-Rhaetian boundary example. Log-likelihood
function using L_4 cannot be approximated closely by
parabola because of central interval with $E^2 = 0$. Mid-
point of this interval and minimum value of parabola
each provide unbiased estimates of mean age; standard
deviation s is overestimated by parabola method.

The approach taken in this paper differs from the one originally
taken by Cox and Dalrymple (1967) as will be discussed in more
detail now. The basic assumptions that the dates are distributed
uniformly through time and subject to measurement errors are
made in both methods of approach. Cox and Dalrymple (1976,
see their fig. 4 on p. 2608) demonstrated that, under these
conditions, the inconsistent dates for younger rocks have
probability of occurrence P_{Iy} with:

$$P_{Iy}(t) = (1/2)\ \mathrm{erfc}\left[\frac{(t - \tau)}{\sigma_m \sqrt{2}}\right]$$

where erfc denotes complementary error function and τ
represents true age of the chronostratigraphic boundary
(boundary between geomagnetic polarity epochs in Cox and
Dalrymple's original paper). The standard deviation for the

measurement errors is written as σ_m. Setting $\tau = 0$ and using the relationship $(1/2)$ erfc $(z/\sqrt{2}) = 1 - \Phi(z)$ it follows that:

$$P_{Iy}(t) = 1 - \Phi\left(\frac{t}{\sigma_m}\right) = f_3\left(\frac{t}{\sigma_m}\right)$$

If t/σ_m is replaced by x, we obtain exactly the weighting function $f_3(x)$ of Figure 1. Consequently, this weighting function can be interpreted as the probability that an inconsistent age t' is measured for younger rocks. Likewise, $P_{Io}(t) = f_3(-t/\sigma_m)$ can be defined for older rocks.

Cox and Dalrymple (1967) next introduced the trial boundary age t_e and defined a measure of dispersion of all inconsistent dates t' with respect to t_e satisfying:

$$D^2(t' - t_e) = \int_{-\infty}^{\infty} (t - t_e)^2 \, P_L(t) \, dt$$

where $P_I(t) = P_{Iy}(t)$ if $t \geq 0$; and $P_{Io}(t)$ if $t \leq 0$. For $t_e = \sigma$, this quantity is a minimum (see Cox and Dalrymple, 1967, fig. 5 on p. 2608). A normalized version of E^2 can be compared directly to the theoretical curve for $D^2(t' - t_e)$ when the number of inconsistent dates is large. This normalization consisted of dividing E^2 by average number of dates per unit time interval. It is noted that $P_I(t)$ does not represent a probability density function, because it can be shown that

$$\int_{-\infty}^{\infty} P_I(t) \, dt = \sqrt{\frac{2}{\pi}} = 0.80 \qquad \text{if } t_e = \tau.$$

In this paper, E^2 is not interpreted as approximately proportional to $D^2(t' - t_e)$. Instead of this, it is regarded as the inverse of a log-likelihood function with Gaussian weighting function. For large samples, good estimates can be obtained using the inconsistent dates only. For small samples, however, significantly better results are obtained by using the consistent dates also and by replacing the Gaussian weighting function by $f_3(x)$.

As pointed out before, all Gaussian weighting functions provide the same mean age of a chronostratigraphic boundary when the maximum likelihood method is used. However, the standard deviation of this mean depends on the selection of the constant in $\exp(-px^2)$. For example, $p = 0.5$ for $f_1(x)$ and $p = 1.0$ for $f_4(x)$ in Figure 1. Assuming that $f_3(x)$ represents the correct weighting function, we can ask for which p the Gaussian function $\exp(-px^2)$ provides the best approximation to $f_3(x)$ with $x \geq 0$. Let u represent the deviation between the two curves, so that $\log_e\{1 - \Phi(x)\} = -px^2 + u$. Minimizing Σu_i^2 for $x_i = 0.1k$ $(k = 1, 2, ..., 20)$ by the method of least squares gives $\hat{p} = 1.13$. Because of the large difference between the two curves near the origin, $\hat{p}$ increases when fewer values x_i are used. It decreases when more values are used. Letting k run to 23 and 24 yields $\hat{p}$ equal to 1.0064 and 0.9740, respectively. These results confirm the conclusion reached before that $f_4(x)$ with $p = 1.0$ provides a better approximation to $f_3(x)$ than $f_1(x)$ with $p = 0.5$.

COMPUTER SIMULATION EXPERIMENTS

A number of computer simulation experiments were performed in order to attempt to answer the following questions: (1) Does the theory of the preceding section (also see Appendix 1) remain valid even when the number of available dates is very small?; (2) How do estimates obtained by the method of fitting a parabola to the log-likelihood function compare to estimates obtained by the method of scoring which is used by statisticians (see e.g. Rao, 1973)?; and (3) How do results derived from the chronograms in GTS compare to those obtained by the maximum likelihood method?

Figure 7 and Table 1 illustrate the first type of computer simulation experiment performed. Twenty-five random numbers were generated on the interval [0, 10]. These numbers with uniform frequency distribution can be regarded as true dates (τ) without measurement errors. The stage boundary was set equal to 5. Values of τ less than 5 belong to the younger stage A, and those greater than 5 to the older stage B (see Table 1). The measurement error was introduced by adding to τ a normal random number with zero mean and standard deviation equal to one. As a result of this, each value of τ was changed into a date t. Some values of t ended up outside the interval [0, 10], such as 11.197 in the first example (Run No. 1 in Table 1 and Fig. 7), and were not used later. In Run No. 1, a single date for the younger state (A) has t > 5, and a date of B has t < 5. Suppose now, for example, that the trial age of the stage boundary t_e is set equal to 4.6, then there are three inconsistent ages for Run No. 1 and these are marked by asterisks in Table 1. Each normalized date x = t - t_e was converted into a z-value by changing its sign if it belongs to the younger stage A. The value of z was transformed into a probability P = Φ(z) for values of t on the interval

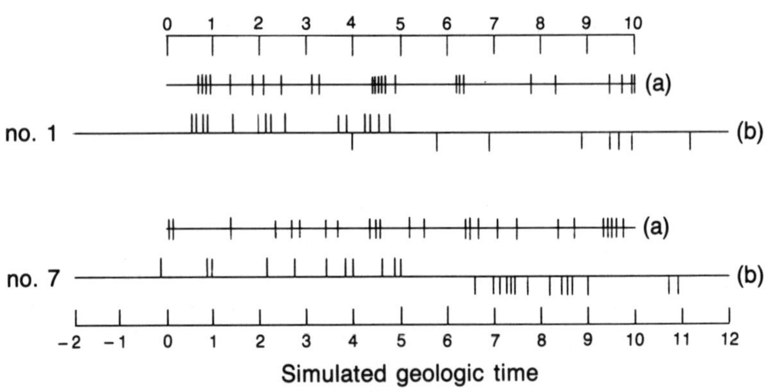

Figure 7. Two examples (Runs No. 1 and No. 7) of computer simulation experiments of first type. True dates (a) were generated first, classified and increased (or decreased) by random amount. Younger and older ages are shown above and below scale (b), respectively.

Table 1. Run 1 for first type of computer simulation experiment.
True dates τ were classified as younger (A) or older (B) than true age of stage boundary (=5). Dates t with measurement error are compared to trial age (t_e = 4.6). Inconsistent ages are indicated by asterisks. $z = -x$ for younger radius (A) and $z = x$ for older rocks (B). Probability P is fractile of standard normal z-value. Total of logs of P gives value of log-likelihood function for t_e = 4.6.

τ		t	x ($= t-4.6$)	z	P	$\log_e P$
4.587	A	4.380	-0.220	0.220	0.5871	-0.5325
7.800	B	8.048	3.448	3.448		
2.124	A	2.193	-2.407	2.407	0.9920	-0.0081
0.668	A	2.239	-2.361	2.361	0.9909	-0.0092
6.225	B	5.802	1.202	1.202	0.8853	-0.1218
9.990	B	9.945	5.345	5.345		
4.896	A	4.574	-0.026	0.026	0.5102	-0.6730
4.606	A*	6.487	1.887	-1.887	0.0296	-3.5211
0.796	A	0.553	-4.047	4.047		
1.855	A	2.526	-2.074	2.074	0.9810	-0.0192
6.292	B	6.923	2.323	2.323	0.9899	-0.0101
3.280	A	1.998	-2.602	2.602	0.9954	-0.0046
2.422	A	1.435	-3.165	3.165		
1.397	A	0.912	-3.688	3.688		
4.538	A	4.365	-0.235	0.235	0.5928	-0.5230
0.830	A	0.803	-3.797	3.797		
6.194	B*	4.033	-0.567	-0.567	0.2854	-1.2540
4.545	A	3.930	-0.670	0.670	0.7490	-0.2890
4.774	A*	4.814	0.214	-0.214	0.4154	-0.8786
0.905	A	0.713	-3.887	3.887		
9.763	B	11.197				
8.285	B	8.902	4.302	4.302		
3.131	A	3.676	-0.924	0.924	0.8224	-0.1955
9.987	B	9.435	4.835	4.835		
9.442	B	9.620	5.020	5.020		

Total = $\overline{-8.0397}$

$[t_e - 3, t_e + 3]$ where $\Phi(z)$ denotes fractile of the normal distribution in standard form. The fractile of 3 is equal to 0.999 with natural logarithm equal to -0.001. For this reason, values outside the interval $t_e \pm 3$ yield probabilities which are approximately 1 (or 0 for log-likelihood function) and these were not used for further analysis. Most probabilities are greater that 0.5. Only inconsistent dates (asterisks in Table 1) give probabilities less than 0.5. The value of the log-likelihood function for t_e is the sum of the logs of the probabilities as illustrated for $t_e = 4.6$ in Table 1.

A second type of computer simulation experiment is illustrated in Table 2. In Table 1, the standard deviation of the normal distribution for measurement errors was kept equal to one for each date. In Table 2, the standard deviation was made a uniform random variable on the interval [0.3, 1.8]. This reflects the fact that in GTS the least precise ages have a standard deviation which is approximately six times as large as that of the most precise ages. The normalized value $x = t - t_e$ now has to be divided by the standard deviation before it can be changed into a z-value with unit variance which can be converted into a probability (see Table 2).

Log-likelihood values for Run No. 1 of the first type of experiment are shown in Table 3 with t_e ranging from 3 to 7 in steps of 0.1. The largest log-likelihood value is reached for $t_e = 5.6$ and this value was selected as the first approximation $\hat{\tau}_{e1}$ of the age of the stage boundary. In total, 21 value of $\hat{\tau}_e$ with $|t_e - \tau_{e1}| \leq 1.0$ were used for fitting a parabola as shown in Figure 8. The fitted parabola is more or less independent of number of values used (=21) and width of neighborhood (=2). However, the neighborhood should not be made too wide because of random fluctuations (local minima or maxima) near $t_e = 3$ or 7 (see e.g. Table 3). These edge effects should be avoided. They are due to the fact that the initial range of simulated time was set arbitrarily equal to 10 in the computer simulation experiments. The peak of this parabola provides the second approximation $m = \hat{\tau}_{e2}$ of the estimated age. The standard deviation (s) of the corresponding normal distribution can be used to estimate the 95 percent confidence interval $m \pm 1.96s$ also shown in Figure 8.

Table 2. Run 1 (see Table 1) repeated using different (randomly selected) measurement errors (σ).

τ		σ	t	x $(= t-4.6)$	x/σ	z	P	$\log_e P$
4.587	A	0.649	4.453	-0.147	-0.227	0.227	0.5898	-0.5279
7.800	B	1.400	8.148	3.548				
2.124	A	1.767	2.245	-2.355	-1.333	1.333	0.9086	-0.0958
0.668	A	0.987	2.219	-2.381	-2.412	2.412	0.9959	-0.0080
6.225	B	0.790	5.891	1.291	1.634	1.634	0.9489	-0.0525
9.990	B	1.403	9.926	5.326				
4.896	A	1.522	4.407	-0.194	-0.127	0.127	0.5969	-0.5969
4.606	A*	1.502	7.432	2.832	1.885	-1.885	0.0297	-3.5166
0.796	A	1.543	0.392	-4.208	-2.802	2.802		
1.855	A	1.662	2.890	-1.710	-1.029	1.029	0.8662	-0.1437
6.292	B	1.666	7.344	2.744	1.647	1.647	0.9502	-0.0511
3.280	A	0.912	2.111	-2.489	2.729	2.729	0.9968	-0.0032
2.422	A	1.451	0.990	-3.610				
1.397	A	0.405	1.201	-3.399				
4.538	A	1.049	4.357	-0.243	-0.232	0.232	0.5916	-0.5249
0.830	A	0.632	0.813	-3.787				
6.194	B*	1.338	3.302	-1.298	-0.970	-0.970	0.1661	-1.7954
4.545	A	1.133	3.846	-0.754	-0.665	0.665	0.7470	-0.2917
4.774	A*	1.304	4.826	0.226	0.173	-0.173	0.4312	-0.8411
0.905	A	1.704	0.578	-4.022				
9.763	B	1.260	11.571					
8.285	B	1.370	9.131	4.531				
3.131	A	1.412	3.900	-0.700	-0.496	0.496	0.6901	-0.3710
9.987	B	1.761	9.015	4.415				
9.442	B	1.667	9.739	5.139				
							Total =	-8.8197

Table 3. Values of log-likelihood functions estimated for Run 1 (first type of computer simulation experiment) and predicted values for parabola fitted by method of least squares. Initial guesses of extreme values are indicated by asterisks.

TIME	LOG-LIKELIHOOD ($\Sigma \log P$)	SUM OF SQUARES (E^2)	PREDICTED LLF	PREDICTED E^2
3.0	-15.58	10.86		
3.1	-14.41	9.37		
3.2	-13.30	8.00		
3.3	-12.27	6.75		
3.4	-11.31	5.63		
3.5	-16.98	13.54		
3.6	-15.83	12.07		
3.7	-14.75	10.73		
3.8	-13.75	9.52		
3.9	-12.81	8.43		
4.0	-11.94	7.46		
4.1	-11.13	6.59		
4.2	-10.39	5.84		
4.3	-9.72	5.21		5.11
4.4	-9.10	4.69		4.69
4.5	-8.54	4.27		4.32
4.6	-8.04	3.93	-7.98	3.99
4.7	-7.59	3.65	-7.57	3.71
4.8	-7.20	3.44	-7.21	3.47
4.9	-6.87	3.27	-6.89	3.28
5.0	-6.58	3.15	-6.61	3.14
5.1	-6.35	3.06	-6.38	3.04
5.2	-6.16	3.02	-6.19	2.99
5.3**	-6.02	3.01**	-6.05	2.98**
5.4	-5.93	3.05	-5.95	3.01
5.5	-5.88	3.13	-5.89	3.09
5.6*	-5.88*	3.24	-5.88*	3.22
5.7	-5.92	3.40	-5.91	3.39
5.8	-6.00	3.59	-5.98	3.61
5.9	-6.13	3.84	-6.10	3.88
6.0	-6.29	4.15	-6.26	4.18
6.1	-6.49	4.51	-6.46	4.54
6.2	-6.73	4.94	-6.71	4.94
6.3	-7.01	5.42	-7.00	5.38
6.4	-7.33	5.97	-7.33	
6.5	-7.69	6.57	-7.71	
6.6	-8.08	7.23	-8.13	
6.7	-8.50	7.91		
6.8	-8.97	8.65		
6.9	-9.47	9.43		
7.0	-10.01	10.24		

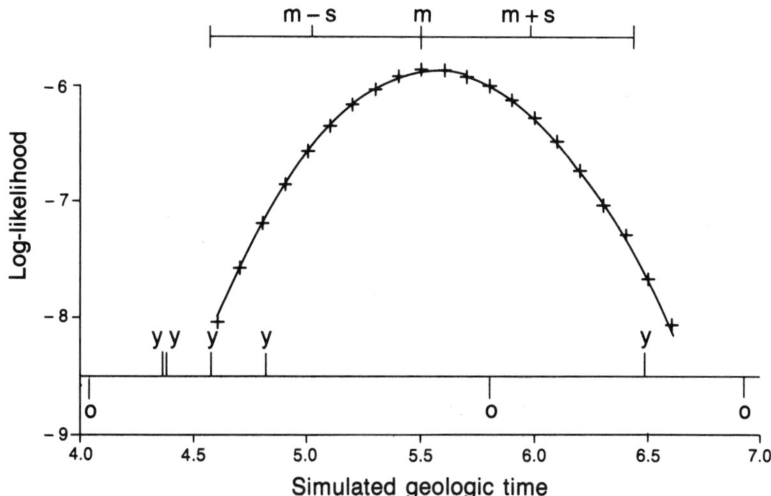

Figure 8.   Maximum-likelihood method (L3) used for estimating mean age of stage boundary in Run 1 (data as in Fig. 7). Standard deviation (s) and 95 percent confidence interval also are shown.

The sum of squares E^2 for L_4, using inconsistent dates only is shown also in Table 3 as a function of t_e. The first approximation of its minimum value is 5.3. The corresponding parabola is shown in Figure 9. The mean age resulting from L_4 is 0.3 less than the mean based on L_3 and its standard deviation is nearly the same. It is fortuitous that the mean based on L_4 is closer to the population mean (=5) than that based on L_3. On the average, the maximum likelihood (L_3) method gives better results (see results for 50 runs given later in this section).

Younger and older ages generated in each of the first 10 (unit variance) computer simulation runs are shown in Figure 10 together with their estimated mean and 95 percent confidence interval using L_3. Theoretically, each population mean (=5) is contained within the 95 percent confidence interval around the sampling mean with a probability of 95 percent. The means and standard deviations used for Figure 10 are listed in Table 4 (maximum likelihood method with parabola). Also listed in Table 4 are corresponding results for L_4 (Gaussian weighting function with parabola). The means based on L_4 are close to those for L_3. The estimated standard deviations tend to be either slightly

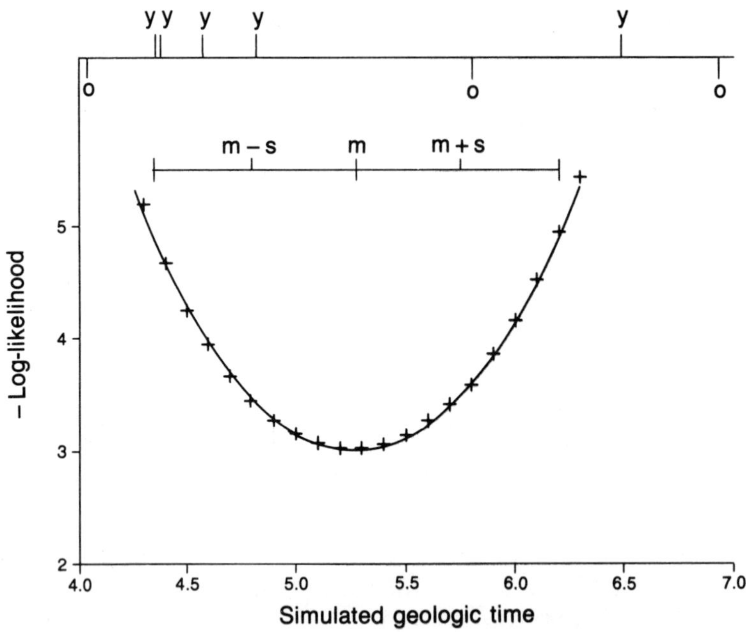

Figure 9.   Chronogram for Run 1 (using L4). Note similarity of s
and 95 percent confidence interval to results shown in
Figure 8.

smaller or much greater. It can be seen from the results for Run
No. 7 shown in Figures 11 and 12 that the greater standard
deviations are due to a break-down of this particular method of
estimation as discussed previously for the Norian-Rhaetian
boundary example (Fig. 6).

Results obtained by the method of scoring (see e.g. Rao, 1973, p.
366-374) also are shown in Table 4. In our application of this
method, the following procedure was followed. As before, the log-
likelihood was calculated for 0.1 increments in t_e and the largest
value of these was used as the initial guess. Suppose that This
value is written as y. Two other values x and z were calculated
representing log-likelihood values close to y at small distances -
10^{-4} and 10^{-4} along the t_e-axis. The quantities $D1 = 0.5(z - x)$ •
10^4 and $D2 = (x - 2y + z)$ • 10^8 were used to obtain a second
approximation of the mean by subtracting D1/D2 from the initial
guess. The procedure was repeated until the difference between
successive approximations became negligibly small. Then the
standard deviation of the estimate is given by $SD = 1/|D2|$.

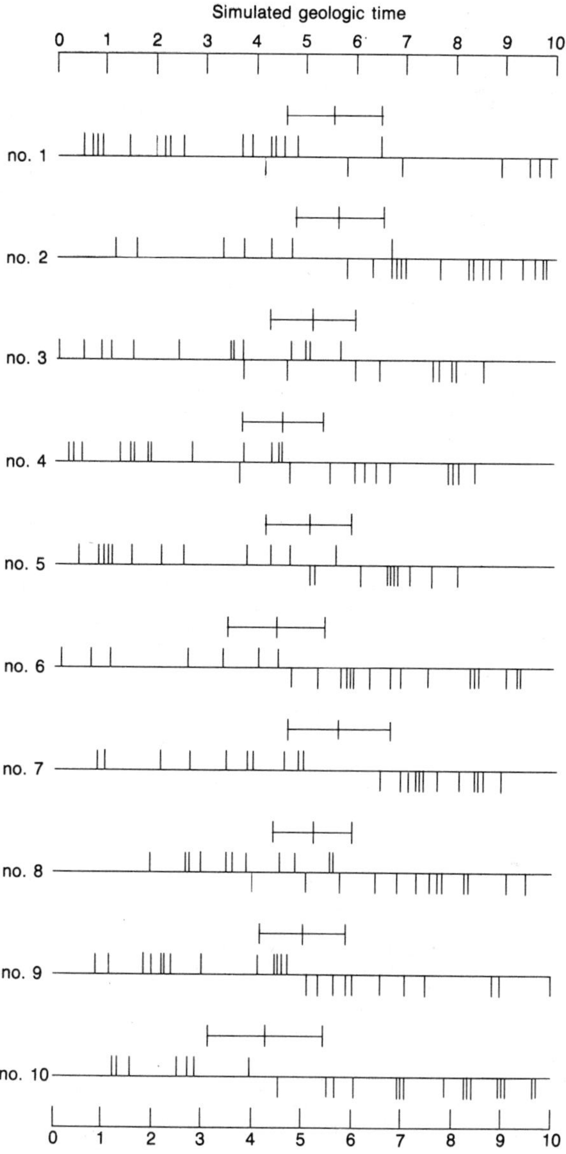

Figure 10. Dates generated in first 10 runs for first type of computer simulation experiments (cf. results for No. 1 and No. 7 shown in Fig. 7). Mean and 95 percent confidence interval estimated by maximum-likelihood method are shown for comparison with true mean (=5).

Table 4. First 10 runs for first type of computer simulation
         experiment. Comparison of results obtained by fitting
         parabola and scoring method, respectively. Standard
         deviations marked by asterisks are too large (cf. Fig. 12).

| Run | | Maximum Likelihood Method | | | | | Gaussian Weighting Function | | | |
| | | Parabola | | Scoring | | | Parabola | | Scoring | |
No.	Mid-point	Mean	S.D.	Mean	S.D.	Mid-point	Mean	S.D.	Mean	S.D.
1	5.6	5.582	0.479	5.554	0.481	5.3	5.269	0.470	5.260	0.500
2	5.7	5.632	0.481	5.663	0.489	6.3	6.190	0.480	6.264	0.500
3	5.1	5.153	0.420	5.142	0.423	4.8	4.884	0.335	4.828	0.316
4	4.5	4.506	0.447	4.507	0.452	4.2	4.321	0.395	4.216	0.354
5	5.1	5.070	0.461	5.089	0.466	5.3	5.217	0.482	5.293	0.408
6	4.4	4.419	0.502	4.448	0.505	4.6	4.625	0.749*		
7	5.7	5.710	0.531	5.728	0.542	5.8	5.767	3.924*		
8	5.2	5.205	0.406	5.200	0.411	5.0	5.025	0.364	5.017	0.408
9	5.0	5.022	0.417	5.018	0.419	5.0	4.966	0.614*		
10	4.2	4.231	0.609	4.232	0.623	4.3	4.248	1.001*		

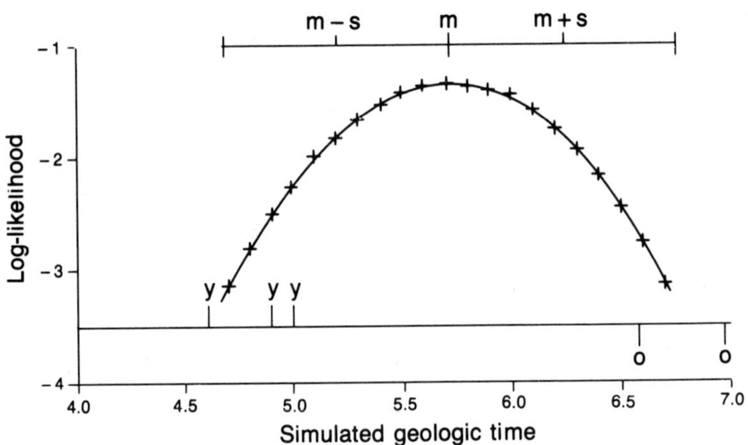

Figure 11. Maximum-likelihood method (L3) used for estimating
           mean age of stage boundary in Run 7 (data as in
           Fig. 7).

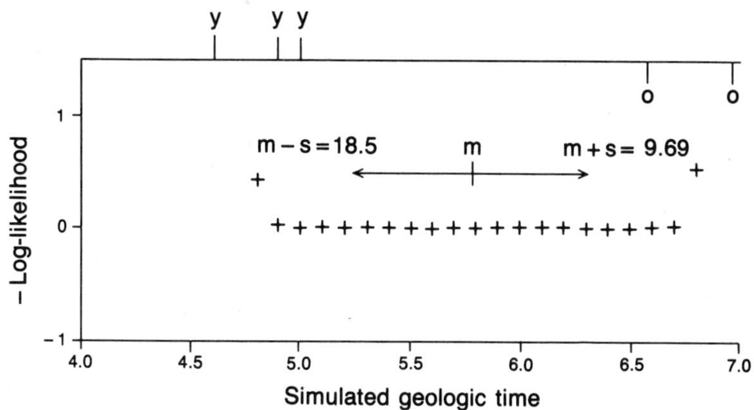

Figure 12. Chronogram for Run 7 using L4. Standard deviation (s) severely over estimates true certainity.

For L_3, the scoring method generally yields estimates of SD which are slightly greater than those resulting from the parabola method. However, the difference is negligibly small (Table 4). For L_4, the scoring method provided an answer in only 6 of the 10 experiments of Table 4.

Table 5 shows similar results for 10 runs for computer simulation experiments of the second type (variable measurement error). Only 11 (instead of 21) log-likelihood values were used for fitting the parabola because initially poor fits occurred in some experiments when the larger neighborhood with 21 values was used. The results in Table 5 are similar to those shown in Table 4. The maximum likelihood method (L_3) provides nearly the same answers for the parabola and scoring methods. The Gaussian weighting function (L_4) failed to provide good estimates of the standard deviations in 5 of the 10 runs (Table 5).

In total, 50 runs were made for each of the two types of experiments. For constant variance of measurement errors, the parabola method for L_3 gave an overall mean equal to 4.9287 and standard deviation 0.4979 as calculated from 50 means. The corresponding numbers for the second type of experiment were 4.9442 and 0.5160. The Gaussian weighting scheme gave overall

Table 5. First 10 runs for second type of computer simulation
experiments. Parabola could not be fitted for mid-points
indicated by double asterisks.

| Run | | Maximum Likelihood Method | | | | Gaussian Weighting Function | | | | |
| | | Parabola | | Scoring | | | Parabola | | Scoring | |
No.	Mid-point	Mean	S.D.	Mean	S.D.	Mid-point	Mean	S.D.	Mean	S.D.
1	5.6	5.583	0.514	5.562	0.534	5.4	5.367	0.500	5.367	0.500
2	4.7	4.712	0.412	4.730	0.423	5.7	5.712	0.333	5.728	0.316
3	4.4	4.352	0.387	4.372	0.404	4.6	4.589	0.846*		
4	5.4	5.401	0.454	5.394	0.472	5.6	5.622	0.358	5.631	0.354
5	5.6	5.620	0.624	5.634	0.630	5.9**				
6	5.1	5.045	0.504	5.058	0.514	4.5	4.512	0.388	4.459	0.408
7	4.4	4.378	0.577	4.375	0.618	4.6	4.470	9.524*		
8	5.8	5.808	0.678	5.837	0.805	5.9**				
9	4.5	4.501	0.532	4.484	0.554	4.0	4.054	0.588*		
10	5.4	5.368	0.350	5.366	0.352	5.1	5.114	0.413	5.092	0.408

means equal to 4.9213 and 4.9414 for the two types, and
corresponding standard deviations equal to 0.5790 and 0.6541,
respectively. If the parabola did not provide a good fit to the
function E^2, because of zero values around its minimum, the mean
was approximated by the mid-point of the range of zero values in
these calculations. The results of the 50 runs for the two
experiments confirm the earlier results described in this section.
Additionally, they show that the Gaussian weighting function
(using L_4) provides results which are almost as good as the
method of maximum likelihood (using L_3).

SMOOTHING WITH THE AID OF CUBIC SPLINE FUNCTIONS

When the ages of a number of successive chronostratigraphic
boundaries have been estimated, they can be improved further by
smoothing with the aid of cubic spline functions. The spline-
fitting technique used here was developed independently by
Reinsch (1967) and Schoenberg (1967). Some of this theory is
reviewed briefly in Appendix 2. The ages shown in Table 6 and
Figure 13 will be used for example. They were derived from
chronograms in GTS with the following relatively minor

Table 6. Ages and estimated standard deviations used for fitting spline-curve No. 1 shown in Figure 13.

Lower boundary of stage	Age	S.D.
1 Maastrichtian (Maa)	72	1.41
2 Campanian (Cmp)	84	1.59
3 Santonian (San)	87.5	1.59
4 Coniacian (Con)	88.5	0.88
5 Turonian (Tur)	91	0.88
6 Cenomanian (Cen)	97.5	0.70
7 Albian (Alb)	113	1.41
8 Aptian (Apt)	} 122	3.18
9 Barremian (Brm)		
10 Hauterivian (Hau)	124	2.83
11 Valanginian (Vlg)	} 135	1.77
12 Berriasian (Ber)		
13 Tithonian (Tth)	145	4.24
14 Kimmeridgian (Kim)	151	2.12
15 Oxfordian (Oxf)	} 158	5.30
16 Callovian (Clv)		
17 Bathonian (Bth)		
18 Bajocian (Baj)		
19 Aalenian (Aal)		
20 Toarcian (Toa)		
21 Pliensbachian (Plb)		
22 Sinemurian (Sin)		
23 Hettangian (Het)	212	4.95
24 Rhaetian (Rht)	213	6.36
25 Norian (Nor)	218	2.83
26 Carnian (Crn)	228	7.78
27 Ladinian (Lad)	238	3.54
28 Anisian (Ans)	} 242	7.43
29 Scythian (Scy)		
30 Tatarian (Tat)	246	7.07
31 Kazanian/Ufimian (Kaz-Ufi)	} 253	8.13
32 Kungurian (Kun)		
33 Artinskian (Art)	268	4.24
Sakmarian/Asselian (Sak-Ass)		

modifications: (a) If the chronograms in GTS for the two boundaries of a stage are the same, indicating absence of dates for that stage, the estimate was assigned to a single point mid-way between the stage boundaries; (b) Imprecise estimates for 6 successive Jurassic stages were not used; (c) When inconsistent dates are missing, the estimated age was set equal to the mid-point of the range for missing data in the chronogram; and (d) The standard deviation was set proportional to the age range listed in the summary time scale (GTS, p. 52-55) with constant proportionality equal to $(1/2)\sqrt{2}$.

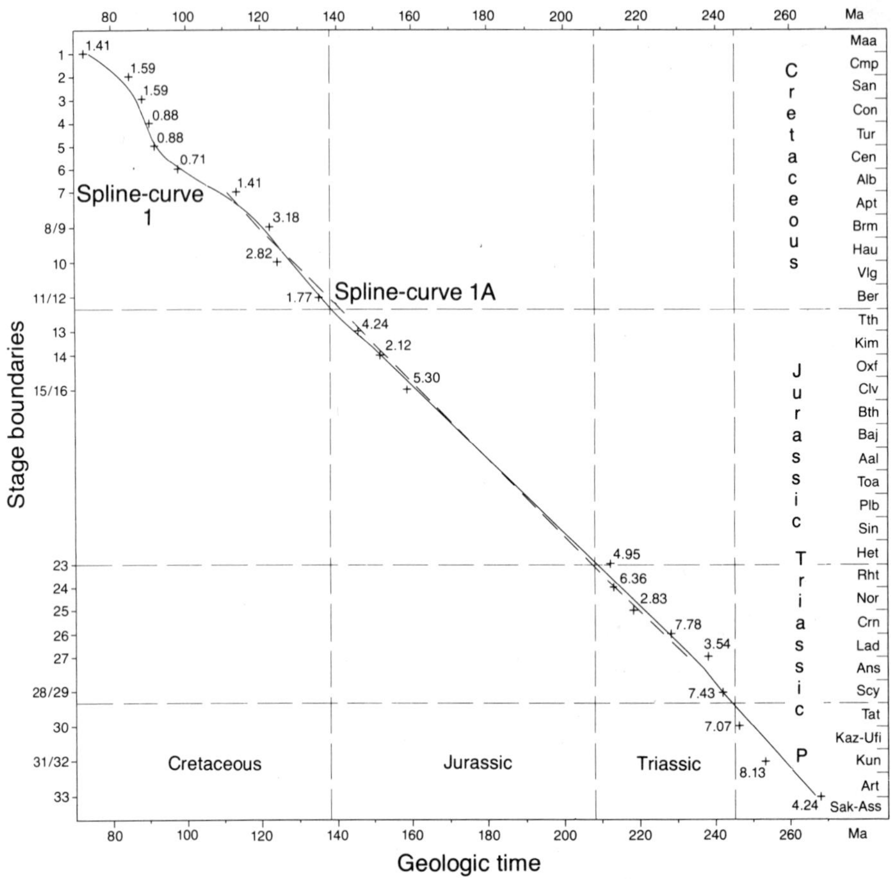

Figure 13. Spline-curves fit to ages of stage boundaries listed in Table 6. Spline-curve 1A was fitted to data for stage boundaries numbered 7 to 27 only.

The latter modification (d) is based on the earlier considerations corroborated by the computer simulation experiments proving that the parabola for L_4 provides a better approximation to the parabola for L_3 than that for L_1. The reason for this can be understood by comparing the weighting functions for these models with one another (see also comparison by method of least squares carried out previously). The combined weight assigned to z-values greater than 1.0 in the Gaussian weighting function for L_1 is about 6 times greater than in that for L_4. In the weighting function for L_3 ($f_3(x)$ in Fig. 1), the weights assigned to z-values greater than 1.0 also are smaller than those for the weighting function of L_1 ($f_1(x)$ in Fig. 1).

A cubic spline-curve was fit to the data in Figure 13 for the following reasons. A spline-curve is smooth because there are no abrupt changes in the rate of change of its slope; the principle of least squares is used; and deviations between observed values (crossed in Fig. 13) and spline-curve are permitted to exist but the sum of squares of these deviations can be regulated. The weight assigned to each observed value is inversely proportional to its variance.

Let the vertical and horizontal axes in Figure 13 represent observations written as x_i, y_i ($i = 1, ..., n$), respectively. Then the smoothing spline-function $f(x)$ to be constructed minimizes $\int f''(x)^2 dx$ among all possible functions $f(x)$ such that

$$\sum_{i=1}^{n} \left[\frac{f(x_i) - y_i}{s(y_i)} \right]^2 \leq S$$

Here the $s(y_i)$ are the standard deviations of the values y_i. The sum of standardized deviations S is a random variable approximately distributed as chi-squared with n degrees of freedom and variance equal to 2n. The expected value of S, which is equal to n, was used in our applications.

It can be seen in Figure 13 that the fitted Spline-curve No. 1 tends to follow the stage boundaries in the Cretaceous more closely because these are relatively precise. In places where the uncertainty is great, the spline-curve tends to become a straight line.

Spline-curve No. 1A shown in Figure 13 was fitted to points for stage boundaries between the Anisian and Cenomanian. It is nearly straight and closely approximates Spline-curve No. 1.

Because the intervals between stage boundaries in the vertical direction of Figure 13 are spaced equally, a straight line in this type of plot would agree with the hypothesis of equal duration of stages. As explained in the introduction, Harland and others

(GTS, 1982) applied linear interpolation between relatively precise stage boundaries defined as tie-points. The boundaries numbered 1 to 7, 27 and 33 were used as tie-points in GTS. Because the crosses for boundaries No. 7 and 27 fall slightly to the right of the fitted spline-curves, the estimates obtained by spline interpolation are younger than those in GTS.

With respect to the Jurassic time scale, Kent and Gradstein (in press) have argued that it is more reasonable to assume equal duration of zones than equal duration of stages. They used Hallam's (1975) ammonite zones for spacing the stage-boundaries in the Jurassic between tie-points at the base of the Kimmeridgian and Hettangian, respectively. On the basis of other evidence including data on rates of seafloor spreading in the Late Jurassic, these authors assumed ages of 156 Ma and 208 Ma for these two stage boundaries (No. 14 and No. 23), respectively.

The values of x_i used for constructing the spline-curve of Figure 13 can be modified by using n_i for number of ammonite zones per stage (see Table 7). The new values x_i shown in Table 7 satisfy

$$x_i = x_{i-1} + cn_i$$
$$x_{12} = 12; \quad i = 13, ..., 23$$

where $c = 11/62 = 0.1774$ represents the ratio of total number of stages (= 11) and zones (= 62) in the Jurassic. The modified spline-curve (No. 2) for equal duration of zones gave results which will be shown later.

The input for spline-curve fitting was modified further by using as tie-points 156 Ma instead of 151 Ma for the Oxfordian-Kimmeridgian and 208 Ma instead of 212 Ma for the Triassic-Jurassic boundary, respectively, setting the standard deviations of these ages equal to zero. As demonstrated in Appendix 2, the spline-curve has the property of passing exactly through points of which the standard deviation is zero. Spline-curve No. 2 with tie-points is shown in Figure 14. The ages of stage boundaries (rounded off to 1 Ma) obtained by three methods of cubic spline-fitting are shown in Figure 15 for comparison with the other age

Table 7. Ages used for fitting spline-curve No. 2 based on equal
         duration of ammonite zones in the Jurassic; without and
         with tie-points, respectively.

i	Stage		n_i	x_i	Age	S.D.
13	Tithonian	(Tth)	8	13.4	145	4.24
14	Kimmeridgian	(Kim)	4	14.1	156	0.00
15	Oxfordian	(Oxf)	7	15.4	158	5.30
16	Callovian	(Clv)	6	16.4		
17	Bathonian	(Bth)	7	17.7		
18	Bajocian	(Baj)	7	18.9		
19	Aalenian	(Aal)	3	19.5		
20	Toarcian	(Toa)	6	20.5		
21	Pliensbachian	(Plb)	5	21.4		
22	Sinemurian	(Sin)	6	22.5		
23	Hettangian	(Het)	3	23.0	208	0.00

estimates. The spline-curves all gave 208 Ma for the age of the
Triassic-Jurassic boundary which is younger than the GTS
estimate of 213 Ma although the same original age determinations
were used.

The spline-curves yield ages of 138 Ma and 140 Ma for the
Jurassic-Cretaceous boundary which are younger than the 144 Ma
age in GTS and Kent and Gradstein (in press). This relatively
young age is due mainly to the effect of (a) a relatively young
Oxfordian glauconite age listed as 148.22 Ma in GTS and as 145 ±
3 Ma in Armstrong (1978) who, in turn, extracted it from Gyji
and McDowell (1970), and (b) four other relatively young
glauconite ages listed in GTS for the Tithonian. If these five dates
would not be used, the spline-curves also would give an age of
approximately 144 Ma for the top of the Jurassic. In the
introduction it was pointed out that Odin (1982) using more
glauconite dates estimated a younger age (130 Ma) for this
boundary.

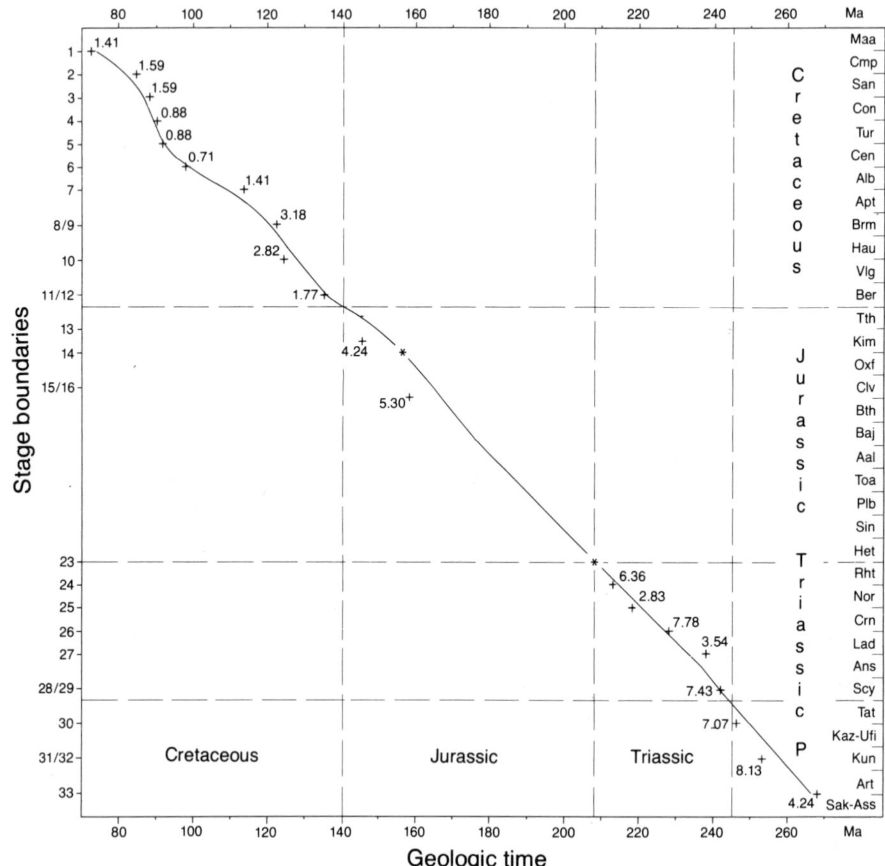

Figure 14. Spline-curve fitted to ages of stage boundaries for Jurassic listed in Table 7. This cubic smoothing spline passes exactly through two tie-points with SD = 0.

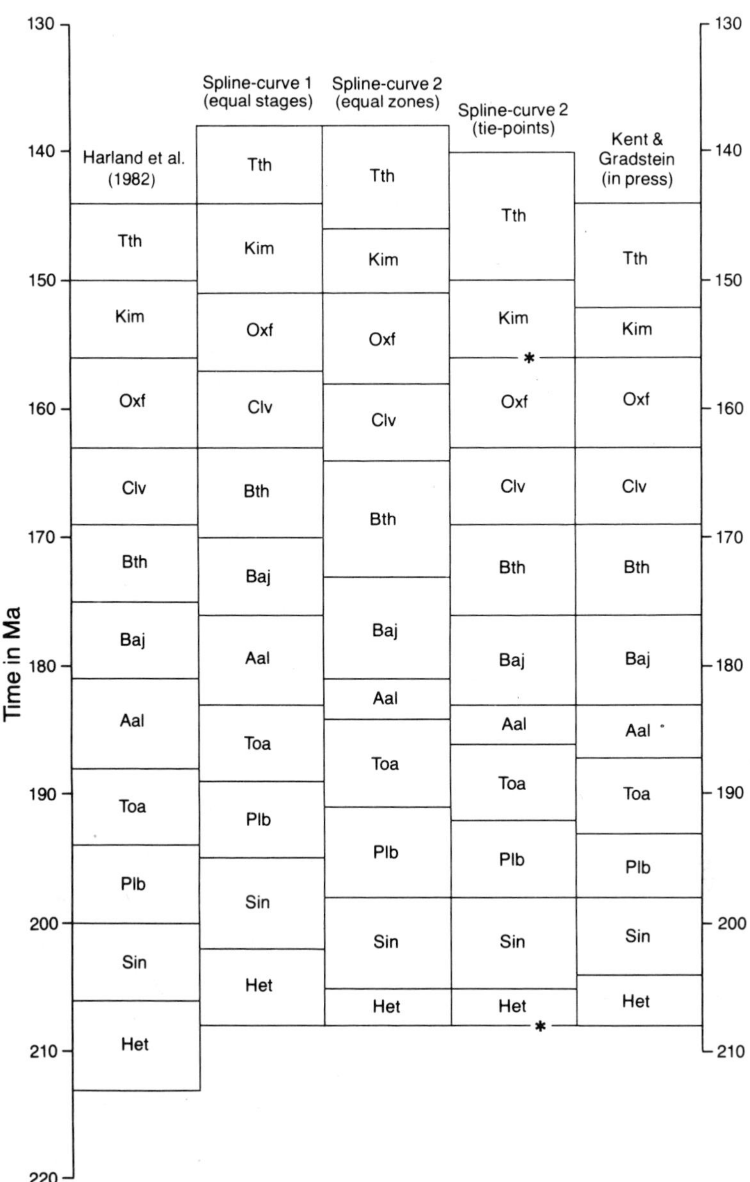

Figure 15. Comparison of spline-curve ages (rounded off to nearest integer Ma value) to ages estimated by Harland and others (GTS, 1982) and by Kent and Gradstein (in press).

CONCLUDING REMARKS

In this paper it has been shown that statistical estimation of the ages of chronostratigraphic boundaries in the geological time table can be improved in two ways: (a) The maximum likelihood method can be used for estimation of the age of individual chronostratigraphic boundaries, and (b) After estimating the ages of a set of successive boundaries by the method of maximum likelihood, these can be improved further by using a cubic spline-curve for smoothing.

ACKNOWLEDGMENTS

Thanks are due to F.M. Gradstein (Atlantic Geoscience Centre, Dartmouth) for valuable suggestions and to S.N. Lew (Geological Survey of Canada, Ottawa) for preparing and running FORTRAN IV programs for the computer simulation experiments. Comments by G.F. Bonham-Carter (Geological Survey of Canada, Ottawa) to improve the text also are gratefully acknowledged.

APPENDIX 1

Asymptotic normality of maximum likelihood estimators

In this appendix, the parameter to be estimated will be written as θ. In our applications, $\theta = t_e$ represents mean age of stage boundary. Following Cramer (1946), Kendall and Stuart (1967, p. 43) show that if (a): the first two derivatives of a likelihood function $L = L(x \mid \theta)$ with respect to θ exist in an interval of θ including the true value θ_0, if (b) :

$$E\left[\frac{\delta \log L}{\delta \theta}\right] = 0 \qquad (1.1)$$

and if (c) :

$$R^2(\theta) = -E\left[\frac{\delta^2 \log L}{\delta \theta^2}\right] = E\left[\left\{\frac{\delta \log L}{\delta \theta}\right\}^2\right] \qquad (1.2)$$

exists and is nonzero for all θ in the interval, the maximum likelihood estimator θ is distributed asymptotically normally with mean θ_0 and variance $1/R^2(\theta)$.

Using Taylor's theorem,

$$\left[\frac{\delta \log L}{\delta \theta}\right]_{\hat\theta} = \left[\frac{\delta \log L}{\delta \theta}\right]_{\hat\theta} + (\hat\theta - \theta_0)\left[\frac{\delta^2 \log L}{\delta \theta^2}\right]_{\theta^*} \qquad (1.3)$$

where θ^* is some value between θ and θ_0. The estimator θ is a root of the likelihood equation

$$\frac{\delta \log L}{\delta \theta} = 0 \qquad\qquad (1.4)$$

so that the left-hand side of Equation (1.3) is zero.

Equation (1.3) may be rewritten as

$$(\hat{\theta} - \theta_0)\, R(\theta_0) \;=\; \frac{\dfrac{\left[\dfrac{\delta \log L}{\delta \theta}\right]_{\theta_0}}{R(\theta_0)}}{\dfrac{\left[\dfrac{\delta^2 \log L}{\delta \theta^2}\right]_{\theta^*}}{\left\{ -R^2(\theta_0) \right\}}} \qquad (1.5)$$

For large samples ($n \to \infty$), $\hat{\theta}$ approaches θ_0 (consistency property of maximum likelihood estimators). Because θ^* lies between θ and θ_0,

$$\left[\frac{\delta^2 \log L}{\delta \theta^2}\right]_{\theta^*}$$

approaches $-R^2(\theta_0)$. The denominator on the right-hand side of Equation (1.5), therefore, approaches 1. The expression

$$\left[\frac{\delta \log L}{\delta \theta}\right]_{\theta_0}$$

can be interpreted as the sum of n independent identical variates

$$\frac{\delta \log f(x \mid \theta_0)}{\delta \theta} \ .$$

The mean and variance of this sum are given by Equations (1.1) and (1.2). The central limit theorem of mathematical statistics therefore applies and the right-hand side is asymptotically normal. It follows that θ also is normal asymptotically with mean θ_0 and variance $1/R^2(\theta_0)$ (Q.E.D.)

It is noted that the topic of properties of maximum likelihood estimators also is reviewed in Rao (1973, p. 364-366) where a cubic term is added to the truncated Taylor series used in Equation (1.3).

APPENDIX 2

Cubic spline-function fitting

Suppose that values $y_i (i = 1, 2, ..., n)$ are available for points x_i with $x_1 < x_2 < ... < x_n$ and that a so-called cubic smoothing spline $f(x)$ is to be fitted according to the model

$$y_i = f(x_i) + e_i \qquad (2.1)$$

where e_i is an error term with known standard deviation s_i. The main purpose of this Appendix is to prove that a cubic smoothing spline can be fitted so that, at one or more tie-points x_k, y_k, $k \in \{1, ..., n\}$ $y_k = f(x_k)$ with $e_k = 0$.

The subject of splines in statistics has been reviewed recently by Wegman and Wright (1983) to which the reader is referred for more background information on splines. The so-called cubic interpolation spline with $e_i = 0$ at all points x_i, y_i $(i = 1, 2, ..., n)$ can be derived as follows. Let $h_i = x_{i+1} - x_i$ for

$$y_i = f(x_i) = a_i + b_i(x - x_i) + c_i(x - x_i)^2 + d_i(x - x_i)^3$$

$$x_i \leq x < x_{i+1} \qquad (2.2)$$

Further,

$$f'(x_i)_- - f'(x_i)_+ = 0 \quad (i = 2, ..., n-1)$$

$$f''(x_i)_- - f''(x_i)_+ = 0 \quad (i = 1, ..., n) \qquad (2.3)$$

with

$$f^{(k)} (k_i)_{\pm} = \lim_{h \to 0} f^{(k)} (x_i \pm h) \quad (k = 0, 1, 2, 3)$$

Continuity of the first two derivatives is assumed. Also,

$$f'' (x_o)_- = f'' (x_n)_+ = 0$$

For the cubic smoothing spline, the condition $e_i = 0$ will be replaced by

$$f (x_i)_- - f (x_i)_+ = 0 \qquad (i = 2, ..., n-1)$$

$$f''' (x_i)_- - f''' (x_i)_+ = 2p \frac{e_i}{s_i} \qquad (i = 1, ..., n) \quad (2.4)$$

where p is a parameter to be determined separately. Then also,

$$f''' (x_o)_- = f''' (x_n)_+ = 0$$

For the cubic interpolation spline, $a_i = y_i$ and

$$a_{i+1} = a_i + b_i h_i + c_i h_i^2 + d_i h_i^3 \qquad (2.5)$$

From the conditions for the second derivative, it follows that

$$c_i = c_n = 0$$

$$d_i = \frac{(c_{i+1} - c_i)}{3h_i} \qquad (i = 1, ..., n-1) \qquad (2.6)$$

Let us now adopt methodology and notation of Reinsch (1967) as follows: $\mathbf{C}$ is a column vector with $(n - 2)$ elements $c_2, \ldots, c_{n-1}$; $\mathbf{Y}$ a column vector with n elements $y_1, \ldots, y_n$; $\mathbf{T}$ a positive definite, tridiagonal matrix of order $(n - 2)$ with $t_{ii} = 2(h_{i-1} + h_i)/3$, $t_{i,i-1} = t_{i+1,i} = h_i/3$; and $\mathbf{Q}$ a tridiagonal matrix with n rows and $(n - 2)$ columns; $q_{i-1,i} = 1/h_{i-1}$, $q_{ii} = -1/h_{i-1} - 1/h_i$, $q_{i+1,i} = 1/h_{i+1}$.

From the continuity of the first derivative, it follows that

$$\mathbf{TC} = \mathbf{Q'Y} \qquad (2.7)$$

In practice, the coefficients of the cubic interpolation spline can be determined by obtaining first the coefficients $c_2, \ldots, c_{n-1}$ from

$$\mathbf{C} = \mathbf{T}^{-1} \mathbf{Q'} \mathbf{Y} \qquad (2.8)$$

then d_i from Equation (2.6); and, finally, b_i from Equation (2.5).

It may be shown that the cubic interpolating spline is the solution of the problem of minimizing mean square curvature or determining the minimum of

$$\int_{-\infty}^{\infty} \{ f''(x) \}^2 \, dx \qquad (2.9)$$

subject to $f(x_i) = y_i$.

The cubic smoothing spline represents the solution of minimizing (2.9) subject to the constraint

$$\sum_{i=1}^{n} \left[\frac{f(x_i) - y_i}{s_i} \right]^2 \leq S \qquad (2.10)$$

where S is a constant. If all standard deviations s_i were known exactly, S would be χ^2-distributed random variable with n degrees of freedom (see text). The expression (2.9) can be minimized subject to (2.10) by use of methods of the calculus of variations (Courant and Hilbert, 1953, chapt. IV). As discussed by Reinsch (1967) a Lagrangian parameter p can be introduced to minimize the functional

$$\int_{-\infty}^{\infty} \{f''(x)\}^2 \, dx \; + \; p \sum_{i=1}^{n} \left[\frac{f(x_i) - y_i}{s_i} \right]^2 \; + \; z^2 \; - \; S \quad (2.11)$$

From the corresponding Euler equations, it follows that f(x) satisfies

$$f'''(x) = 0, \quad x_i < x < x_{i+1} \qquad (i = 1, ..., n-1) \quad (2.12)$$

with boundary conditions as in Equations (2.3) and (2.4). Equation (2.12) represents a cubic spline function. Suppose the **A** is a column vector with n elements a_i, and **D** represents a diagonal matrix of order n with $d_{ii} = s_i$.

Continuity of the first derivative now results in

$$\mathbf{TC} = \mathbf{Q'A} \qquad (2.13)$$

and the condition for the third derivative in Equation (2.4) yields

$$\mathbf{QC} = p\mathbf{D}^{-2}(\mathbf{Y} - \mathbf{A}) \qquad (2.14)$$

Equations (2.5) and (2.6) remain valid. From Equations (2.13) and (2.14), it follows that

$$\mathbf{C} = p \, (\mathbf{Q'} \, \mathbf{D}^2 \, \mathbf{Q} + p\mathbf{T})^{-1} \, \mathbf{Q'} \, \mathbf{Y} \qquad (2.15)$$

and

$$\mathbf{A} = \mathbf{Y} - p^{-1} \mathbf{D}^2 \mathbf{Q} \, \mathbf{C} \qquad (2.16)$$

The coefficients of the cubic smoothing spline are obtained by solving Equations (2.15), (2.16), (2.6), and (2.5) respectively, provided that p is known (see later). Suppose now that the point x_k, y_k, $k \in \{1, ..., n\}$ is a tie-point which is known with certainty so that $s_k = 0$. Because the kth row of $\mathbf{D}$ then consists of zeros only, it follows from Equation (16) that $a_k = y_k$. This, in turn, indicates that the cubic smoothing spline has $e_k = 0$ and passes exactly through the tie-point (Q.E.D.).

Equation (2.11) has to be minimized with respect to z and p leading to

$$pz = 0 \qquad (2.17)$$

$$\sum_{i=1}^{n} \left[\frac{f(x_i) - y_i}{s_i} \right]^2 = S - z^2 \qquad (2.18)$$

From Equations (2.15) and (2.16) it follows that

$$S - z^2 = \mathbf{DQ} \, (\mathbf{Q'} \, \mathbf{D}^2 \mathbf{Q} + p\mathbf{T})^{-1} \, \mathbf{Q'} \, \mathbf{Y} \qquad (2.19)$$

From Equation (2.17) it may be concluded that either $p = 0$ or $z = 0$. Because the right-hand side of Equation (2.19) is nonnegative, $p = 0$ is possible only if $S \geq s^2$. This represents the situation that the cubic smoothing spline is reduced to a straight line. In this limit ($p = 0$, $S \to z^2$, $z \to 0$), the best fitting line of least squares arises with

$$a_i = \left\{ 1 - \mathbf{D}^2 \, \mathbf{Q} \, (\mathbf{D'} \, \mathbf{D}^2 \, \mathbf{Q})^{-1} \right\} y_i$$

$$b_i = \frac{(a_{i+1} - a_i)}{(x_{i+1} - x_i)} \qquad (2.20)$$

$$c_i = d_i = 0$$

If $z = 0$, p must be determined from Equation (2.19) with its left-hand side reduced to S. Algorithms for this were provided originally by Reinsch (1967 and 1971).

It is noted that minimizing expression (2.9) subject to (2.10) is one of two equivalent methods for obtaining the cubic smoothing spline. The spline solution of Equations (2.5), (2.6), (2.15), and (2.16) also minimizes

$$\sum_{i=1}^{n} \left[\frac{y_i - f(x_i)}{s_i} \right]^2 \qquad (2.21)$$

subject to

$$\int_{-\infty}^{\infty} \{f''(x)\}^2 \, dx \leq g_0 \qquad (2.22)$$

where g_0 is a constant (cf. Wegman and Wright, 1983, p. 356). This latter condition establishes connection with the so-called least-squares regression splines. The second approach also was taken by Watson (1984) who showed that spline approximations can be regarded as special solutions of Kriging problems.

REFERENCES

Armstrong, R. L., 1978, Pre-Cenozoic Phanerozoic time scale, in Cohee, G. V., Glaessner, M. G., and Hedberg, H. D., eds., Contributions to the geologic time scale: Am. Assoc. Petroleum Geologists, Studies in Geology No. 6, p. 73-91.

Cox, A. V., and Dalrymple, G. B., 1967, Statistical analysis of geomagnetic reversal data and the precision of potassium-argon dating: Jour. Geophysical Research, v. 72, no. 10, p. 2603-2614.

Courant, R., and Hilbert, D., 1953, Methods of mathematical physics, vol. 1: Wiley-Interscience, New York, 561 p.

Cramer, H., 1946, Mathematical methods of statistics: Princeton Univ. Press, Princeton, New Jersey, 575 p.

Gyji, R. A., and McDowell, F. W., 1970, Potassium argon ages of glauconites from a biochronologically dated Upper Jurassic sequence of northern Switzerland: Eclogae Geol. Helvetiae, v. 63, no. 1, p. 11-118.

Hallam, A., 1975, Jurassic environments: Cambridge Univ. Press, Cambridge, 269 p.

Harland, W. B., 1983, More time scales: Geol. Magazine, v. 120, no. 4, p. 393-400.

Harland, W. B., Cox, A. V., Llewellyn, P. G., Pickton, C. A. G., Smith, A. G., and Walters, R., 1982, A geologic time scale: Cambridge Univ. Press, Cambridge, 131 p.

Kendall, M. G., and Stuart, A., 1961, The advanced theory of statistics, vol. 2: Hafner, New York, 676 p.

Kent, D. V., and Gradstein, F. M., in press, A Jurassic to Recent chronology, in Tucholke, B. E., and Vogt, P. R., eds., The geology of North America: Geol. Soc. America.

Odin, G. S., ed., 1982, Numerical dating in stratigraphy, parts I and II: Wiley-Interscience, Chichester, 1040 p.

Rao, C. R., 1973, Linear statistical inference and its applications: John Wiley & Sons, New York, 625 p.

Reinsch, C. H., 1967, Smoothing by spline functions: Numerische Mathematik, v. 10, p. 177-183.

Reinsch, C. H., 1971, Smoothing by spline functions. II: Numerische Mathematik, v. 16, p. 451-454.

Schoenberg, I. J., 1967, On spline functions, in Shisha, O., ed., Inequalities: Academic Press, New York, p. 255-291.

Watson, G. S., 1984, Smoothing and interpolation by kriging and with splines: Jour. Math. Geology, v. 16, no. 6, p. 601-615.

Wegman, E. J., and Wright, I. W., 1983, Splines in statistics: Jour. Am. Statist. Assoc., v. 78, no. 382, p. 351-365.

BIASED KRIGING: A THEORETICAL DEVELOPMENT

Carol Spease and James R. Carr

University of Missouri, Rolla

ABSTRACT

Kriging was developed to be a best linear unbiased estimator using
a theoretical development to assure a minimum variance of
estimation error. The Lagrangian function which assures this
minimization constrained such that the weights (λ) sum to one
(unbiasedness) is

$$L\left(\lambda_i , \mu\right) = \sigma^2 - 2\sum_i \lambda_i\,\sigma\left(x_o\,x_i\right)$$

$$+ \sum_i \sum_j \lambda_i\,\lambda_j\,\sigma\left(x_o\,x_j\right) - 2\mu\left(\sum \lambda_i - 1\right) \qquad (A)$$

In (A), biasedness can be introduced by changing $(\Sigma\lambda\text{-}1)$ to $(\Sigma\lambda\text{-}N)$,
where N is the new sum of weights. Yet, differentiating either
equation with respect to λ and μ results in
formula

$$\sum_i \sum_j \lambda_i\,\sigma\left(x_i\,x_j\right) - \mu = \sum_i \sigma\left(x_o\,x_i\right) \qquad (B)$$

Hence, the same kriging system is used except N is introduced in
the right-hand vector instead of 1. This allows each covariance

value, σ, in (B) to be computed using a variogram, as with unbiased kriging. Biased kriging is useful for favoring a particular portion of a histogram. By allowing the sum of weights to be greater than one, as an example, the high end of the histogram can be favored.

INTRODUCTION

By using kriging as an unbiased estimator (Matheron, 1963), maximum accuracy in data estimation is achieved. That is, accuracy measured as a mean square error between estimated and actual data values is a minimum using kriging as an unbiased estimator. With kriging, the expected value of each estimate is the data mean. By describing an estimator as being unbiased indicates that, through the estimation process, the original mean of the data is unchanged. With unbiased estimation, however, whereas data values near the mean are unchanged, data values having a low frequency of occurrence in a sample are lost in favor of the data values near the mean.

This result is shown in Figure 1. In using kriging, the original histogram of a sample is changed. During unbiased estimation, the observed frequency of values near the mean increases, whereas values whose observed frequencies are low, such as those comprising the tails of a normal distribution, are lost.

For some applications, loss of low-frequency data is not a severe problem. In circumstances where the initial histogram of the data shows a low dispersion about the mean, the loss of low frequency data values will not be significant. Furthermore, for applications where overall estimation accuracy is of utmost importance, unbiased estimation is the most accurate technique to use for data estimation. This is verified in Figure 2.

In practice, there are some applications where low-frequency data are of utmost importance for interpretation. Two examples illustrate this importance. In the field of seismology, for example, observations of high levels of ground motion have greatest implications for earthquake resistant design. If high-level ground motion is underestimated in favor of the mean ground-motion level, this could result in the underdesign of buildings for earthquake resistance.

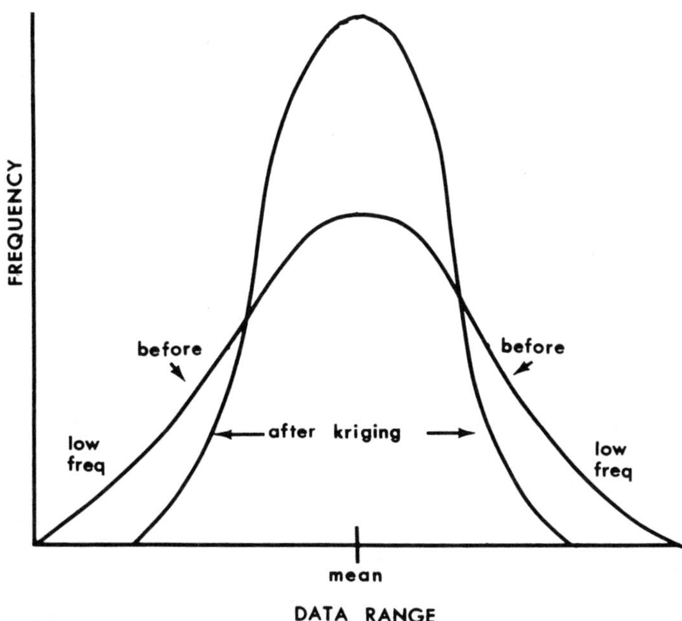

Figure 1. Histograms. Kriging is seen to result in histogram of lower dispersion and higher frequency of mean values relative to original histogram.

As another example, foundation design depends on knowledge of soil density. A measure of soil density is the N-value, measured by driving a split spoon sampler into soil with a 64 kg hammer. The N-value measures the number of hammer blows required to drive the hammer 30 cm. Low N-values are most critical for foundation design because they indicate weak or loose soils. Through unbiased estimation, however, low N-values usually are overestimated in favor of mean N-values.

For these examples, it would be useful to modify kriging to allow this procedure to be sensitive to any range of a histogram. This would de-emphasize the importance of mean values for some applications.

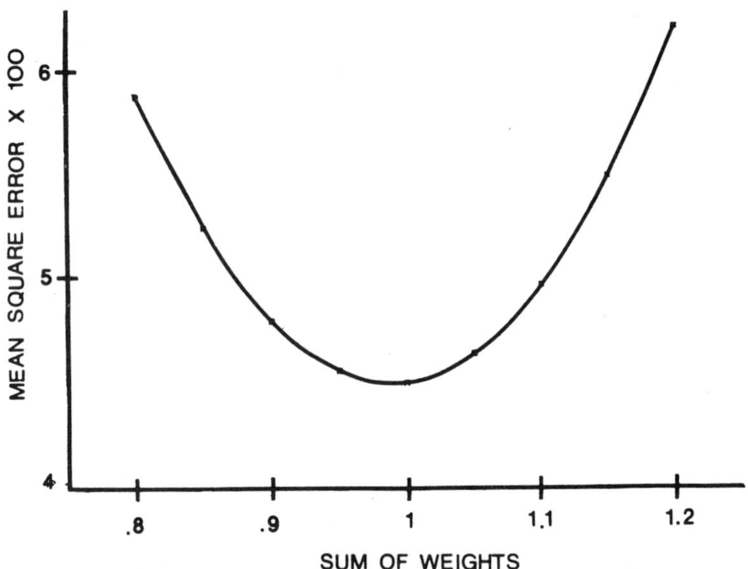

Figure 2. For estimation of given random function, greatest accuracy is achieved for unbiased estimation (where sum of weights is 1.00).

Introduction of Bias Into Kriging

Kriging, a linear estimator, has a simple mathematical form. Expressed as an equation, kriging is (Matheron, 1963)

$$Z^{*}(x_o) = \sum_{i=1}^{N} \lambda_i Z(x_i) \qquad (1)$$

where $Z(x_i)$ is a realization of a stationary random function, Z, at location, i, and $Z^{*}(x_o)$ is the estimate of the value of the random function, A, at location, o.

In Equation 1, the weights, λ, are assigned relative to the spatial structure shown by observed realizations of Z. The objective of kriging is to minimize the variance of the error of estimation subject to the constraint of unbiasedness. This, therefore, is a constrained minimization procedure. To achieve unbiasedness, the weights, λ, are constrained to sum to unity.

In mathematics, a Lagrangian function is used to maximize, or minimize, an outcome subject to constraints. For kriging, the Lagrangian function used to assure a minimum estimation variance whereas placing a constraint on the weights, λ, is (Knudsen and Kim, 1977):

$$L(\lambda_i, m) = \sigma^2 - 2\sum_i \lambda_i \sigma(x_o x_i) + \sum_i \sum_j \lambda_i \lambda_j \sigma(x_i x_j) - 2\mu(\sum_i \lambda_i - 1), \qquad (2)$$

where μ is the Lagrangian multiplier, σ^2 is the sample variance, $\sigma(x_o x_i)$ is the covariance between the estimation location and known data locations and $\sigma(x_i x_j)$ is the covariance between known data locations.

In Equation 2, unbiasedness is assured by the term $2\mu(\Sigma\lambda_i - 1)$. If Equation 2 is differentiated with respect to λ and μ, the resulting system of equation is

$$\sum_i \sum_j \lambda_j \sigma(x_i x_j) - \mu = \sum_i \sigma(x_i x_j) \qquad (3)$$

This system of equations is used to solve for λ and μ.

A bias can be introduced into the kriging system by replacing $2\mu(\Sigma\lambda_i - 1)$ with $2\mu(\Sigma\lambda_i - N)$ in Equation 2. Here, N, is the new sum of the weights. To favor data values greater than the mean, N is set greater than 1; to favor data values less than the mean, N is set less than 1.

By making this change in Equation 2 and differentiating again, Equation 3 results. In other words, the same system of equations is used to solve for λ regardless of the constraint placed on λ.

To illustrate this, the general matrix form of Equation 3 is shown for unbiased estimation:

$$
\begin{bmatrix} & & & 1 \\ & & & \cdot \\ (\sigma(x_i x_j)) & & & \cdot \\ & & & \cdot \\ & & & \cdot \\ & & & 1 \\ 1 \ldots 1 & 0 & 0 \end{bmatrix}
\begin{bmatrix} \lambda_1 \\ \lambda_2 \\ \cdot \\ \cdot \\ \cdot \\ \lambda_N \\ \mu \end{bmatrix}
=
\begin{bmatrix} (\sigma(x_0 x_i)) \\ \\ \\ \\ 1 \end{bmatrix}
$$

For biased estimation, the matrix system of Equation 4 is changed only slightly to become:

$$
\begin{bmatrix} & & & 1 \\ & & & \cdot \\ (\sigma(x_i x_j)) & & & \cdot \\ & & & \cdot \\ & & & \cdot \\ & & & 1 \\ 1 \ldots 1 & 0 & 0 \end{bmatrix}
\begin{bmatrix} \lambda_1 \\ \lambda_2 \\ \cdot \\ \cdot \\ \cdot \\ \lambda_N \\ \mu \end{bmatrix}
=
\begin{bmatrix} (\sigma(x_0 x_i)) \\ \\ \\ \\ N \end{bmatrix}
$$

where the only change is made in the right-hand measurement vector where the sum of the weights is changed from 1 to N.

Despite the change in the constraint placed on λ, these weights reflect the spatial relationship displayed by observed realizations of the random function, Z. Intersample covariance, as well as the covariance between the location at which an estimate is made and known data locations, is used for estimation. Moreover, covariance values are not modified to introduce bias into kriging. Hence, a variogram is useful for biased kriging.

Applications of Biased Kriging

Biased kriging, as mentioned previously, is warranted to assure proper estimation of critical data values. For these values, unless they are approximately equal to the mean of the data, a weighting sum of other than unity is required to optimize the estimation of these values.

An example is peak acceleration values recorded during the 1971 San Fernando, California (USA) earthquake (Trifunac and others, 1973). The mean of these data is approximately 60 cm/sec/sec, yet 25% of these instrument recordings showed peak acceleration to be greater than 90 cm/sec/sec. Using unbiased kriging, these high-level ground motion values tended to be underestimated grossly and the average mean square error over all estimation was approximately
900 cm^2/sec^4.

To demonstrate the advantages of biased kriging, nine values of high-peak acceleration recordings were isolated in a cross-validation procedure. These nine values are listed in Table 1. Also tabulated is the estimation result for each of the nine instrument values as a function of bias. For these particular data, optimum estimation occurs for a weighting sum equal to 1.45. Optimization of results is assessed on the basis of average square error of estimation.

Of further interest, the average error differs linearly with the bias. For a weighting sum of 1.45, the average error is approximately zero. Coincidentally, the average square error is a minimum where the average error is zero. This also is true for unbiased kriging. Figure 3 documents the linear variation of the error with respect to bias and also verifies that the average square error is a minimum for an average null error. This also is apparent from Table 1.

Peak acceleration data were useful in showing the utility of biased kriging for estimation of values greater than the mean. Yet, this procedure can be used to estimate data values less than the mean as well. A good example is afforded by N-values useful for foundation design. The mean of these data was 15. For this example, the objective was to optimize estimation for N values less than 15. This optimization was assessed through cross validation to compute the average square error of estimation.

112

Table 1. Peak acceleration data, 1971 San Fernando, California (USA) earthquake, observation greater than 90 cm/sec/sec.

Sum of Weights

Actual	0.90 Est	0.90 Diff²	0.95 Est	0.95 Diff²	1.00 Est	1.00 Diff²	1.05 Est	1.05 Diff²	1.10 Est	1.10 Diff²	1.15 Est	1.15 Diff²	1.20 Est	1.20 Diff²	1.40* Est	1.40* Diff²	1.45* Est	1.45* Diff²	1.50* Est	1.50* Diff²
132	70	3844	75	3249	79	2809	82	2500	85	2209	88	1936	92	1600	108	576	112	400	116	256
126	73	2809	77	2401	81	2025	85	1681	89	1369	92	1156	96	900	112	196	116	100	120	36
95	87	64	91	16	95	0	99	16	102	49	106	121	110	225	126	961	130	1225	134	1521
100	89	121	92	64	96	16	100	0	104	16	107	49	111	121	127	729	131	961	135	1225
106	86	400	90	256	94	144	97	81	101	25	105	1	109	9	125	361	129	529	133	729
120	74	2116	77	1849	80	1600	83	1369	86	1156	89	961	91	784	108	144	112	64	116	16
148	57	8281	60	7744	63	7225	66	6724	69	6241	72	5776	75	5329	91	3249	95	2809	99	2401
116	64	2704	67	2401	71	2025	74	1764	77	1521	81	1225	84	1024	100	256	104	144	108	64
97	85	144	89	64	93	16	97	0	101	16	104	49	108	121	124	729	128	961	132	1225
1/9 Σ Diff²	2276		2004		1762		1570		1400		1253		1124		800		799		830	
1/9 Σ Diff	- 355		- 322		- 290		- 258		- 226		- 195		- 163		- 36		- 4		+ 28	

*Extrapolated from the trend shown by the results for sum of weights 0.90 through 1.20.

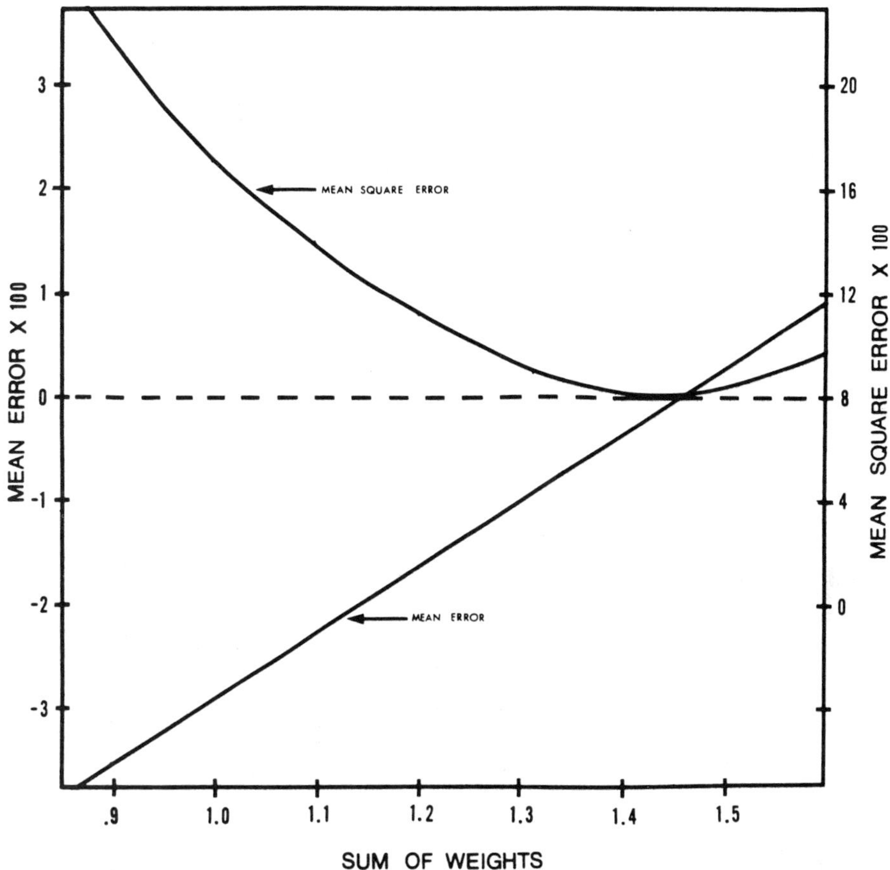

Figure 3. By isolating high levels of peak acceleration, average error is seen to change linearly with bias. Moreover, average square error is minimum where average error is zero.

Twenty N-values comprised that portion of the sample of N-values having a value less than the mean. These data are listed in Table 2. Also tabulated is the cross-validation result for each data value with respect to bias. For these data, an optimum weighting sum is 0.75. At this weighting sum, the average error is zero.

CONCLUSION

Much attention is given to the concept of data stationarity in the field of geostatistics. To be sure, unbiased, linear estimation in

Table 2. N-values for sand, obervations less than 15.

| | Sum of weights | | | | | | | | | |
| | 0.65 | | 0.70 | | 0.75 | | 0.80 | | 0.85 | |
Actual	Est	Diff^2	Est	Diff^2	Est	Diff^2	Est	Diff^2	Est	Diff^2
12	14	4	15	9	16	16	16	16	17	25
10	11	1	12	4	12	4	13	9	14	16
14	10	16	11	9	11	9	12	4	13	1
11	10	1	10	1	11	0	12	1	13	4
12	13	1	14	4	15	9	15	9	16	16
13	12	1	13	0	13	0	14	1	15	4
7	15	64	16	81	16	81	17	100	18	121
13	9	16	10	9	11	4	12	1	12	1
14	8	36	9	25	10	16	10	16	11	9
12	10	4	11	1	12	0	13	1	13	1
11	10	1	11	0	11	0	12	1	13	4
7	5	4	6	1	6	1	7	0	8	1
11	6	25	7	16	7	16	8	9	9	4
12	6	36	7	25	7	25	8	16	9	9
7	6	1	7	0	8	1	9	4	9	4
9	8	1	8	1	9	0	10	1	11	4
14	7	49	8	36	9	25	9	25	10	16
7	8	1	9	4	10	9	10	9	11	16
8	8	0	8	0	9	1	10	4	11	9
7	7	0	7	0	8	1	8	1	9	4
1/20 Σ Diff^2		13.1		11.3		10.9		11.4		13.5
1/20 Σ Diff		−29		−15		0		14		29

the form of kriging is meant to be applied to data possessing a strong stationarity. That is, the mean of the data is invariant spatially and no trend in the data is apparent. The accuracy of kriging may be diminished by nonstationary data. Here, a trend in the data invalidates the expectation that each estimate is the mean of the data.

Nonstationary data displays a regular change in the local data mean in a specific spatial direction. Biased kriging was <u>not</u> developed to correct for this type of data. Nor did the development of biased kriging grow out of research using nonstationary data. The data used for the examples presented in Tables 1 and 2 are second-order stationary. To be specific, any data, stationary or otherwise, comprises a histogram. Biased kriging, as herein presented, has the objective of optimizing estimation for specific ranges of a histogram. The concept of the stationarity of data has nothing to do with this objective.

ACKNOWLEDGMENTS

Results presented in Tables 1 and 2 were obtained using the University of Missouri, Rolla Computer Center. The reference for the sand data is proprietary.

REFERENCES

Knudsen, H.P., and Kim, Y.C., 1977, A short course on geostatistical ore reserve estimation: Dept. Mining and Geol. Engineering, Univ. Arizona, 224 p.

Matheron, G., 1963, Principles of geostatistics: Econ. Geology, v. 58, no. 8, p. 1246-1266.

Trifunac, M.D., and others, 1973, Strong motion earthquake accelerograms, digitized and plotted data, v. II. Corrected accelerograms and integrated ground velocity and displacement curves, pts C-S: California Inst. Technology, prepared for the National Science Foundation, February 1973, distributed by the National Technical Information Service, United States Department of Commerce, 2000 p.

KRIGING HYDROCHEMICAL DATA

Donald E. Myers

University of Arizona

ABSTRACT

As a part of the National Uranium Resource Evaluation Program (NURE) water samples were collected from existing wells in all the continental United States. These samples were analyzed for some 30 elements and ions. Data were assembled for each 2 degrees RMTS quadrangle. The objectives of the NURE program included identification of areas favorable for exploration and producing estimates of recoverable resources. Other authors have reported on the use of pattern recognition, cluster analysis, and discriminant analysis to identify favorable areas.

In cooperation with the Uranium Resource Evaluation Group at Oakridge, the author utilized data from Plainview Quadrangle (Plainview, Texas) to examine the effectiveness of kriging to contour data on 13 variables including uranium. These variables were selected for their chemical association with the deposition or leaching of uranium salts. Because of strong dissimilarities between the Ogallala (Pliocene) and Permian groupings, the data were segregated.

Variograms were computed for each variable, separately for the Permian and Ogallala. Variogram models were cross-validated using randomly selected data subsets. In addition to kriged contour maps for the 13 variables and kriging variance maps in both the Permian and Ogallala, weighted linear sums also were considered. Two different weightings were considered, the weights were determined by a discriminant analysis model. Unusual regions were identified as those for which the kriging error exceeded two kriging standard deviations. These regions were correlated strongly with those identified by a discriminant analysis model and by the quadrangle evaluation.

INTRODUCTION

The objective of the National Uranium Resource Evaluation (NURE) Program was "to provide a systematic appraisal of the uranium resources of the conterminous United States and Alaska" Everhart (1977). It was envisioned that geologic, radiometric, hydrogeochemical, and stream-sediment data would be collected and analyzed systemically in an appropriate manner and evaluations prepared for various geographical regions. The Hydrogeochemical and Stream Sediment Reconnaissance (HSSR) Program was one facet of the data collection process. As indicated by Roach (1978) it was expected that statistical analysis would play an important role but the types of techniques to be used were not specified. Kane (1977) has described the application of standard statistical techniques such as cluster and factor analysis to HSSR data. This paper will present the results of applying kriging to hydrogeochemical data from the Plainview (Texas) Quadrangle (NTMS).

Kriging is a linear estimation technique that incorporates the spatial dependence of the variable in question. Kriging was investigated as a tool to delineate geochemical patterns, identify anomalous areas, and dispersion properties of hydrogeochemical variables in a quantitative way.

THE PLAINVIEW QUADRANGLE AND HSSR DATA

As a part of the HSSR Program, water samples were obtained from approximately 900 wells in the Plainview Quadrangle which then were analyzed at the ORGD analytical laboratory. Each sample location was identified by latitude and longitude and observed values recorded for some thirty hydrogeochemical variables. A complete listing is given in the open-file quadrangle report (URE, 1978) and the data also are available.

A detailed discussion of the geology of the Plainview Quadrangle and the reasons for its selection for this study is contained in Myers and others (1980). Briefly, the reasons include the following: (1) relative simplicity of the geology, there being only two major geologic formations; (2) the Plainview Quadrangle was the only one for which the quadrangle evaluation was complete (Amaral, 1979); and (3) good overall groundwater sample coverage. There are 473 sites in the Permian units and 375 in the Ogallala Formation. Data for twelve variables were considered in the Permian and thirteen in the Ogallala. These will be listed later.

KRIGING

The statistical technique known as kriging was developed by Matheron (1965, 1971, 1973) and his associates at the Centre de Geostatistique, ENSMP, France to provide an improved method of ore-grade estimation. It also has been used as a contouring technique in hydrology and more recently for soil mapping (Journel and Huijbrechts, 1978; Burgess and Webster, 1980). The application to hydrogeochemical data reported here apparently is new and utilizes kriging for more than just contouring.

The reader is referred to Journel and Huijbrechts (1978) or Myers and others (1980) for a more complete derivation of the kriging estimator and its properties; the following is a brief summary.

Let x be a geographical position and $z(x)$ the value of a hydrogeochemical variable such as uranium concentration at x. $z(x)$ can be considered as a function defined on a two-dimensional region, well and aquifer depths were not used. If the form of the function were known it would be sufficient to substitute simply the coordinates for x and compute $z(x)$. $z(x)$ is in general an irregular function and its form is not known; only the values $z(x_1), \ldots ,z(x_n)$ at sample locations $x_1, x_2, \ldots ,x_n$. The problem then is to estimate or predict the value at an unsampled location. Inverse Distance Weighing (IDW) and Polygonal are two widely used methods both of which incorporate local influences. The parallel study of the Plainview Quadrangle data using IDW is reported in Kane and others (1982) and is also in Myers and others (1980). Trend-Surface Analysis (TSA) attempts to fit a smooth function to the data and does not incorporate local influences. To derive the kriging estimator it is assumed that $z(x)$ is a realization of a random function $Z(x)$. It then is necessary to determine appropriate statistical characteristics of $Z(x)$ to proceed with estimation. Matheron determined that two conditions were sufficient

$$E[Z(x) - Z(x+h)] = 0 \qquad (1)$$

$$\text{for all } x, h \ (h \text{ a vector})$$

$$\text{Var}[Z(x) - Z(x+h)] = 2\gamma(h) \qquad (2)$$

where $\gamma(h)$ depends only on h. $\gamma(h)$ quantifies the spatial dependence. Equation (1) implies the absence of drift. If $\gamma(h) = \gamma(|h|)$, $|h| = $ length of h, $Z(x)$ is said to be isotropic. The kriging estimator is of the form

$$Z^*(x) = \sum_{j=1}^{n} \lambda_j(x)\, Z(x_j) \qquad (3)$$

where the λ_j 's are selected so that Z* is an unbiased estimator, that is

$$E[Z^*(x) - Z(x)] = 0 \qquad (4)$$

and the variance of the error is minimal

$$\text{Var}[Z^*(x) - Z(x)] = \sigma_K^2(x) \qquad (5)$$

The minimal value σ_K^2, is termed the kriging variance. The λ_j's are obtained from the linear system

$$
\begin{bmatrix}
\gamma_{11} & \cdots & \gamma_{1n} & 1 \\
 & & & \\
 & \cdot & & \\
 & & \cdot & \\
\gamma_{n1} & \cdots & \gamma_{nn} & 1 \\
 & & & \\
1 & & 1 & 0
\end{bmatrix}
\begin{bmatrix}
\lambda_1 \\
\cdot \\
\cdot \\
\lambda_n \\
\mu
\end{bmatrix}
=
\begin{bmatrix}
\gamma_1 \\
\cdot \\
\cdot \\
\gamma_n \\
1
\end{bmatrix}
\qquad (6)
$$

where $\gamma_{ij} = \gamma(x_1 - x_j)$, $\gamma_i = \gamma(x - x_i)$ and μ is a Lagrange multiplier
introduced to solve the minimization problem. To apply kriging it
is necessary to test whether conditions (1) and (2) are satisfied
and to determine $\gamma(h)$ which is termed the variogram. $\gamma(h)$ can be
estimated by the sample variogram

$$
\gamma^{*}(h) = \frac{1}{2N} \sum (z(x + h) - z(x))^2 \qquad (7)
$$

where N is the number of pairs of sample locations at "distance" h
and the sum is over all such pairs. It is known that - $\gamma(h)$ must be
conditionally positive definite and the usual procedure is to try to
fit $\gamma^{*}(h)$ to one of several known standard functional types. For
ore-grade estimation there is a moderate amount of accumulated
experience which provides guidance on selecting a functional
form for $\gamma(h)$. For example, if a spherical model is used for $\gamma(h)$,
then the parameters are related in a direct way to the graph of
$\gamma^{*}(h)$. Because this was a new application there are no references
to previous studies. It was determined that the sample
variograms also provided insight into the hydrogeochemical
groupings.

SAMPLE VARIOGRAMS

Because the HSSR data were collected for the purpose of aiding in
the assessment of uranium resources the principal variable of
interest was uranium, the other variables were selected because of

their usefulness in identifying or predicting uranium occurrences. Sample variograms (svg) were computed and plotted for the following elements or variables: Uranium, Boron, Barium, Calcium, Lithium, Magnesium, Molybdenum, Sodium, Vanadium, Sulfate, Specific Conductance, Total Alkalinity, and Arsenic (Ogallala only). Because of the small number of sample locations in the Dockum Group these were omitted. The svg's were computed and plotted separately for the Permian and the Ogallala. To test whether an isotropic model for γ could be used, directional plots also were made. Because the sample locations were not on a uniform grid, few pairs had the same distance although the total number of pairs is large. For plotting purposes, N was taken to be 1000 and the plotted value was an average. Because it is general practice to fit geochemical data to a log normal distribution svg's also were computed and plotted for logarithmic transformed data.

Figure 1 shows the plots for uranium, directional and isotropic, for the Ogallala Formation. Figure 2 shows the same plots for the Permian. Figure 3 illustrates the possible contrast between the Ogallala and Permian units. Figure 4 illustrates how the svg differentiates between variables.

As described in Myers and others (1983) the variograms were of four graphical types.

It is of interest to note that all of the plots exhibited a "nugget effect". A complete set of the svg plots and a description of the computer program is presented in Myers and others (1980)

FITTING THE SEMIVARIOGRAMS

After computing and plotting the svg's these plots were used to determine whether anisotropic models must be used, the type of functional forms that might be appropriate, and the range of influence. A subjective decision was made to use only isotropic models and the functional forms to be used were determined to be power, logarithmic, and exponential. It was assumed that the range of dependency would be considerably less than 90 miles and in fact functions were fitted only on the first 30 miles, it is a general characteristic of the kriging estimator that the coefficients, for locations far away, will be small. As yet there are no adequate statistical tests for determining the best model. Having identified an appropriate functional form for each variable

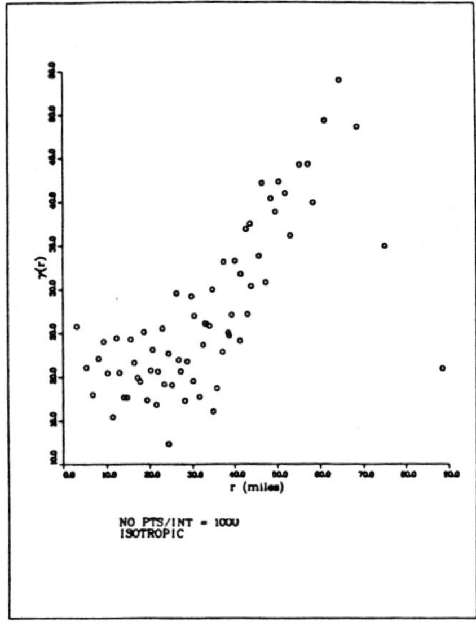

a. Isotropic - Untransformed Data

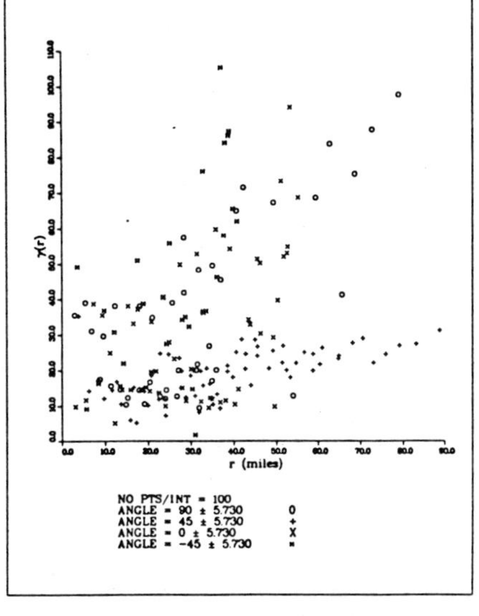

b. Anistropic - Untransformed Data

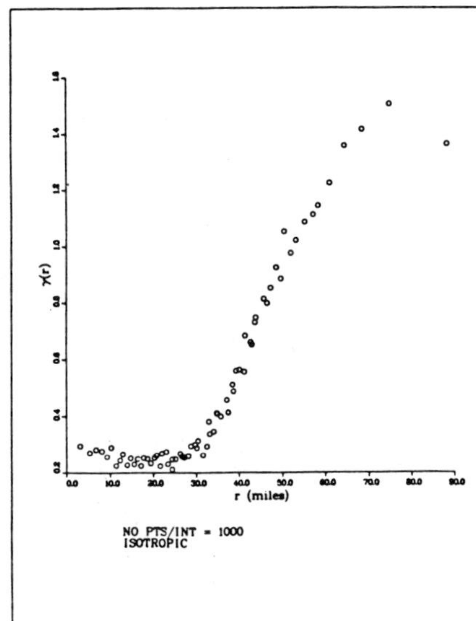

c. Isotropic - Log Transformed Data

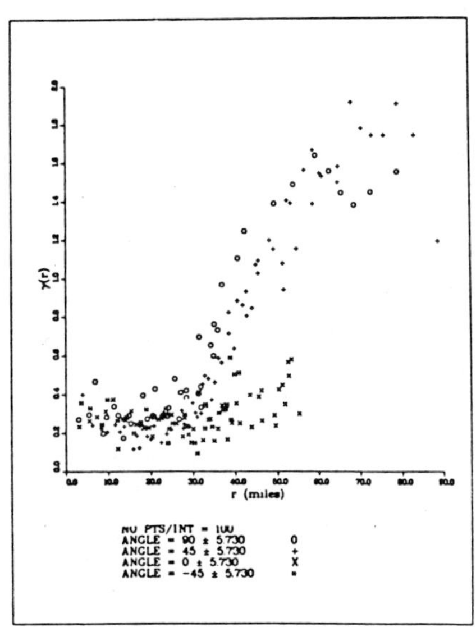

d. Anistropic - Log Transformed Data

Figure 1. Uranium semivariograms for Ogallala Formation.

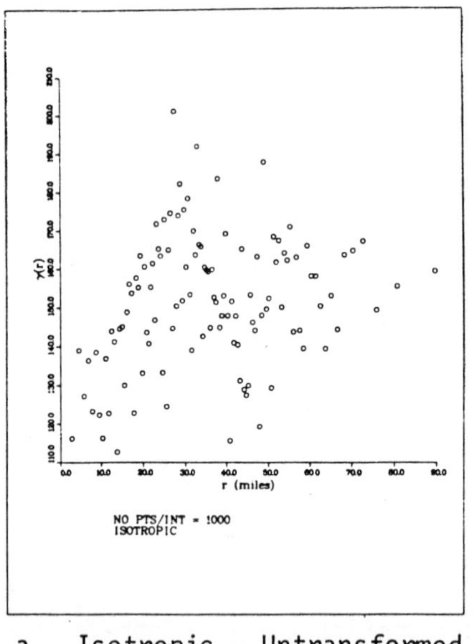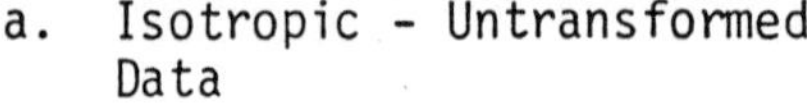

a. Isotropic - Untransformed
    Data

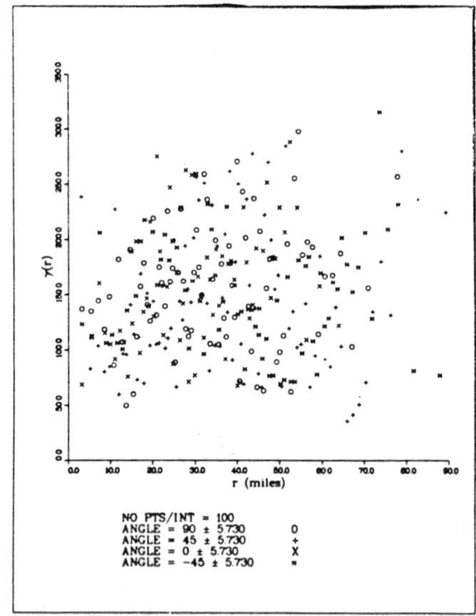

b. Anistropic - Untransformed
    Data

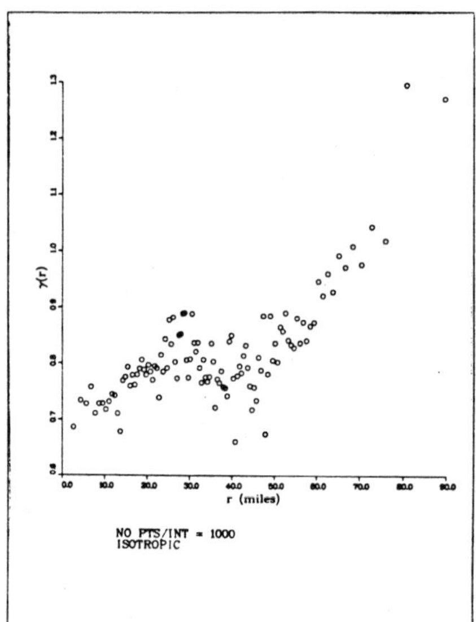

c. Isotropic - Log Trans-
    formed Data

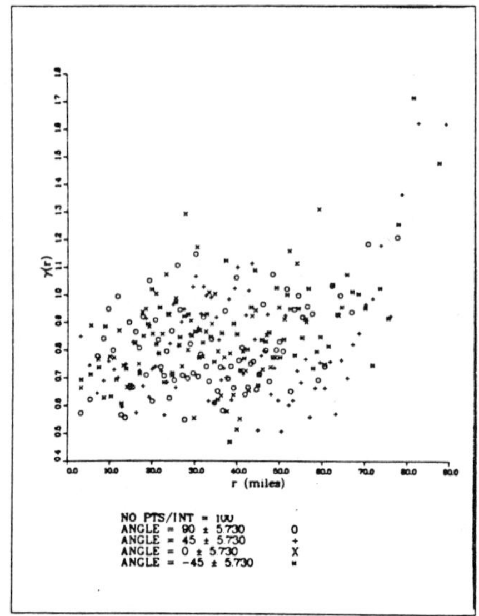

d. Anistropic - Log Trans-
    formed Data

Figure 2. Uranium semivariograms for Permian units.

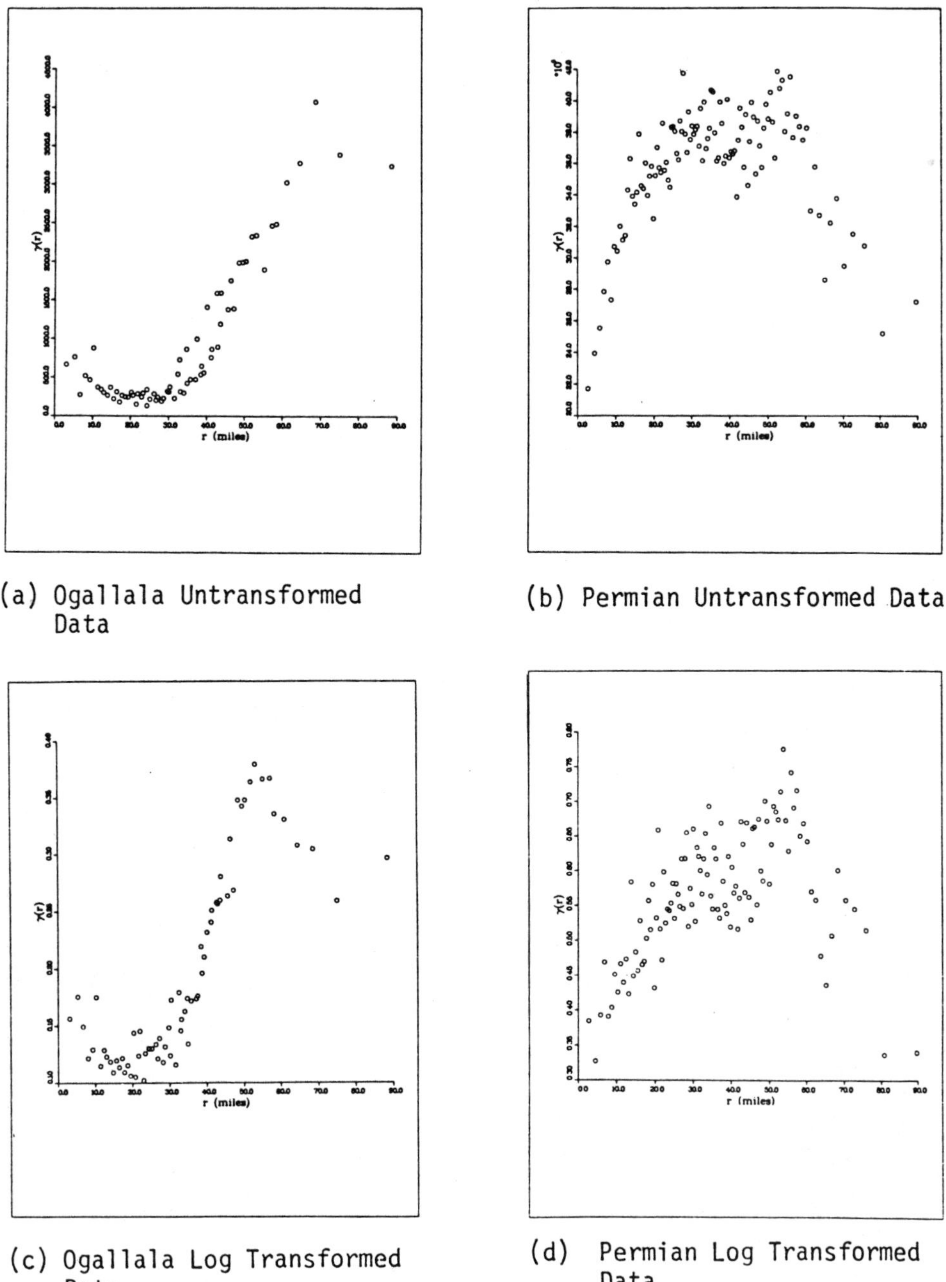

(a) Ogallala Untransformed
Data

(b) Permian Untransformed Data

(c) Ogallala Log Transformed
Data

(d) Permian Log Transformed
Data

Figure 3. Calcium isotropic semivariograms for Ogallala Formation
and Permian units.

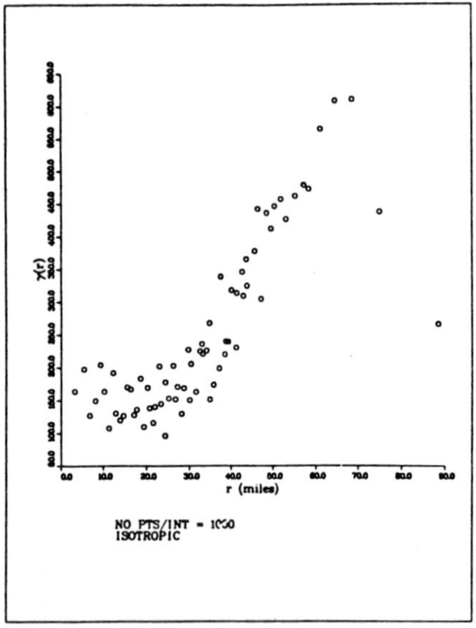

a. Isotropic - Untransformed
    Data

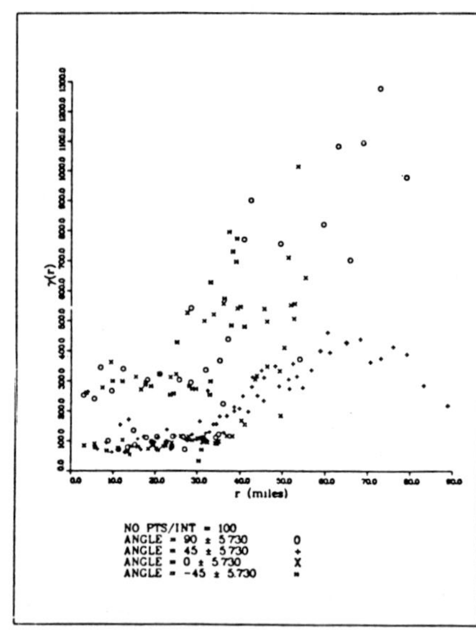

b. Anistropic - Untransformed
    Data

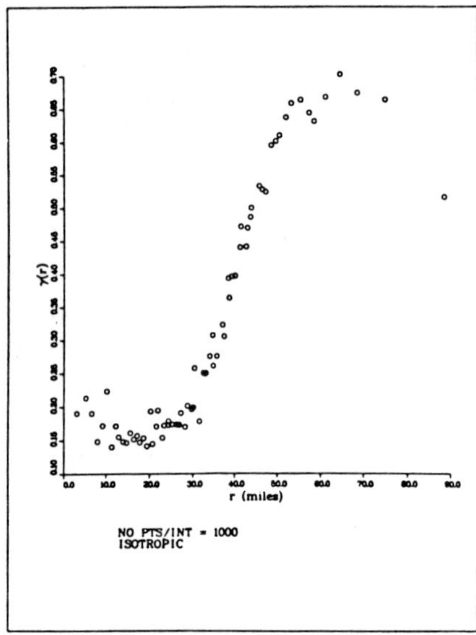

c. Isotropic - Log Trans-
    formed Data

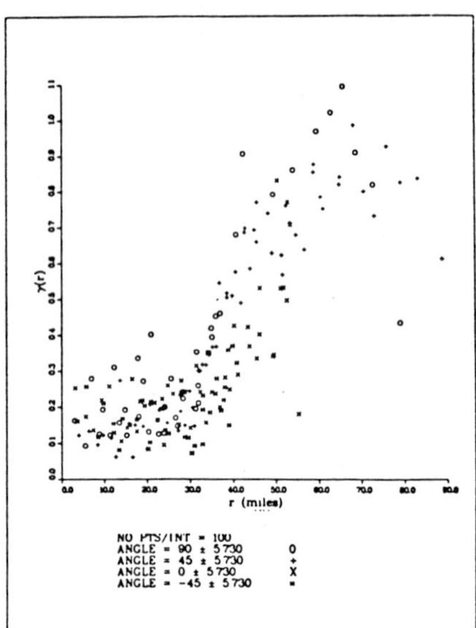

d. Anistropic - Log Trans-
    formed Data

Figure 4-1. Magnesium semivariograms for Ogallala Formation.

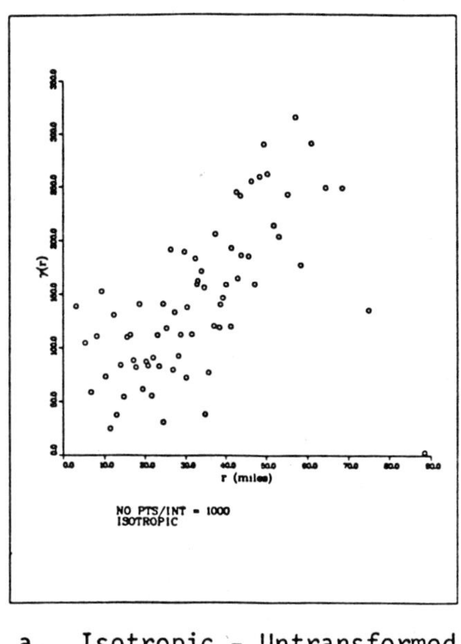

a. Isotropic - Untransformed
    Data

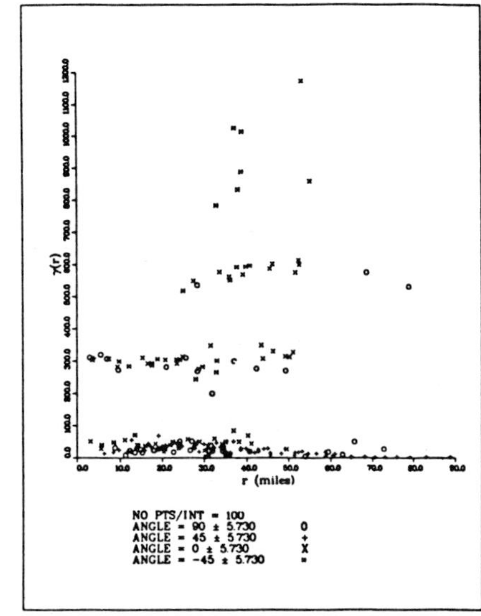

b. Anistropic - Untransformed
    Data

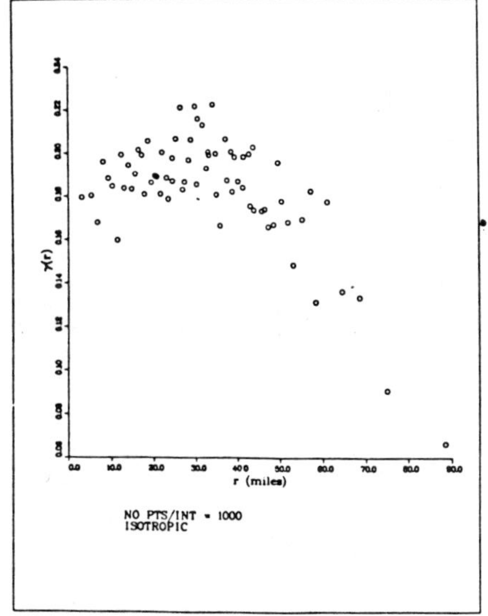

c. Isotropic - Log Trans-
    formed Data

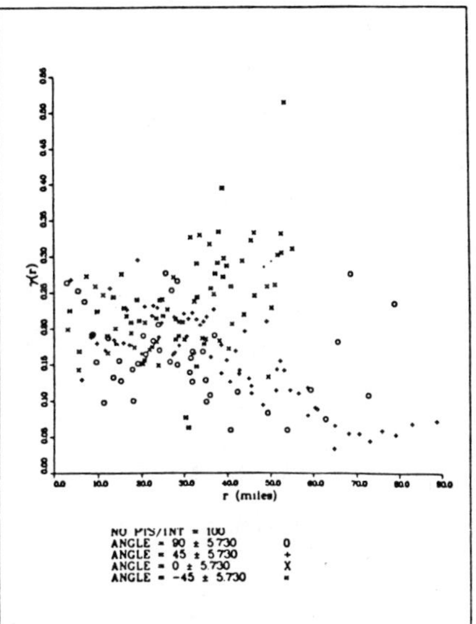

d. Anistropic - Log Trans-
    formed Data

Figure 4-2. Molybdenum semivariograms for Ogallala Formation.

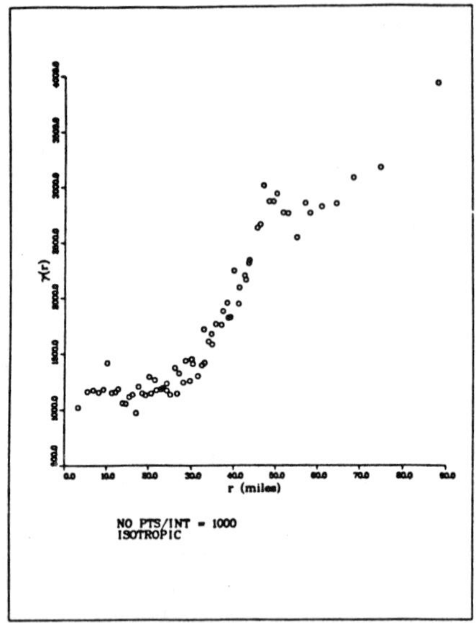

a. Isotropic - Untransformed Data

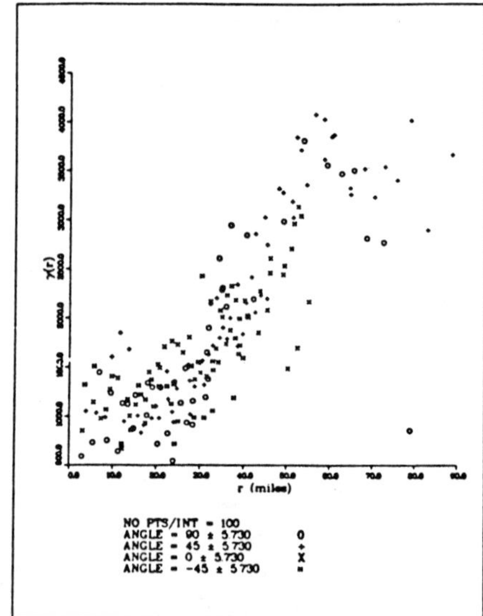

b. Anistropic - Untransformed Data

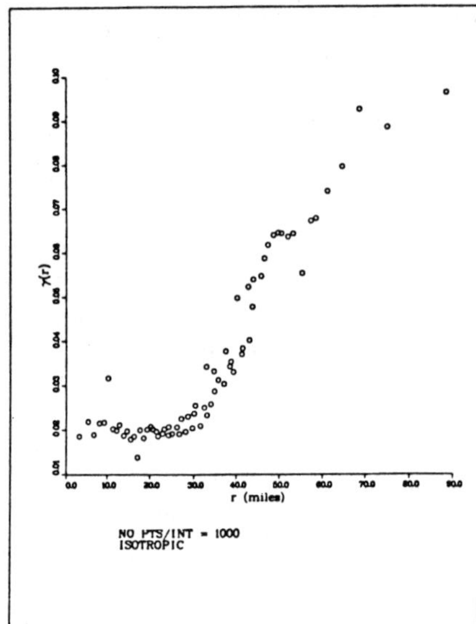

c. Isotropic - Log Transformed Data

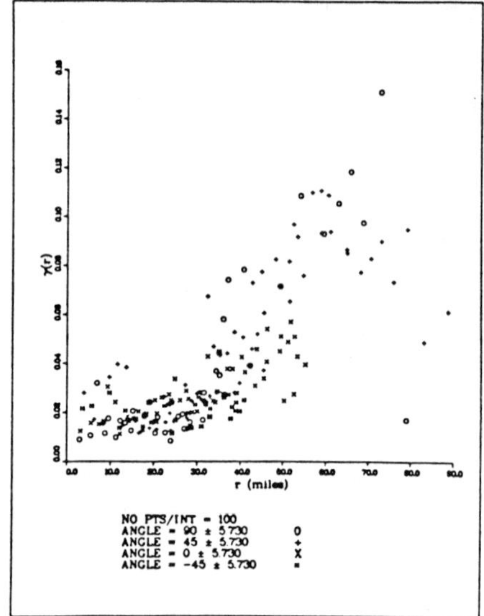

d. Anistropic - Log Transformed Data

Figure 4-3. Total alkalinity semivariograms for Ogallala Formation.

and in each formation, determination of exponents or coefficients was made by standard nonlinear regression analysis. This is not necessarily defensible mathematically but does provide a practical way to fit functions to the plots. For some variables more than one form was considered and for most variables both the transformed and untransformed data were used.

Because kriging is an exact interpolation technique, an estimate can only be obtained for a sample location by deleting that location from the data set. The difference between the observed and the estimated values is indicative not only of how well the selected function form fits but also how plausible are the mathematical assumptions, of isotropy, stationarity, and zero drift. These residuals normalized by the kriging standard deviation can be used to identify unusual locations.

To compute these residuals for all sample locations and all proposed models would have required large amounts of computer time. In order to select the final functional forms to be used, only one for each variable, a random subsample of 100 locations in the Ogallala and 100 in the Permian were selected. For each of these a kriged estimate was obtained using only other sample locations within 10 miles. This was done for each proposed functional model for each variable.

A good fit for the semivariogram should produce collectively small residuals, except that large normalized residuals might indicate anomalies.

Three statistics were computed for each functional model. These were

$$S = \frac{\sum_{i=1}^{100} (z_i - z_i^*)^2}{\sum_{i=1}^{100} (z_i - \bar{z})^2} \qquad (8)$$

Table 1. S, S^*, $\bar{S}$ Values

	S	S*	$\bar{S}$	Unit
U	2.033	0.105	1.330	Ogallala
L-U	0.933	1.354	0.888	Permian

$$S^* = \frac{\sum_{i=1}^{100} (z_i - z_i^*)^2 / \sigma_i^2}{\sum_{i=1}^{100} (z_i - \bar{z})^2 / s_i^2} \qquad (9)$$

$$\bar{S} = \frac{\sum_{i \varepsilon D} (z_i - z_1^*)^2}{\sum_{i \varepsilon D} (z_i - \bar{z})^2} \qquad (10)$$

z_i represents the observed value at location x_i, z_i^* the kriged estimated and z the sample mean. σ_i^2 is the estimation variance for location x_i and D= $\{i \ / \ |z_i - z_i^*| \ / \ \sigma_i < 2\}$. The denominators are almost the sample variance. S, S*, $\bar{S}$ provide a comparison between the kriging estimator and the sample mean as an estimator. Table 1 tabulates the values for S, S*, $\bar{S}$ for Uranium (Ogallala) and Log-Uranium (Permian) for the models that were used subsequently (for the 100 test locations).

The use of transformed data does introduce a bias, that is when a nonlinear transformation such as the logarithm is used the kriging estimator is nonlinear and in general not an unbiased estimator. Journel and Huijbrechts (1978) suggest ways to remove this bias but this was not incorporated in the preliminary study.

INTERPRETATION OF PLAINVIEW DATA

One of the ways that kriging can be used to interpret geochemical data is to produce contour plots. This was done for all 13 variables in the Ogallala and the 12 in the Permian. Those for uranium have been combined into one plot as shown in Figure 5. By overlaying these plots on the plot of favorable areas as determined by Amaral (1979), it is seen that there is strong coincidence of high concentration contours with areas A and B in the Ogallala.

There is some coincidence with area D in the Permian but it seems that area D should be extended southwest. In the NURE quadrangle report correlations are tabulated for each pair of variables. Those showing the highest correlation with uranium, magnesium, vanadium, lithium, total alkilinity. and arsenic (Ogallala only) do not exhibit similar contour patterns as uranium. Of perhaps equal interest are the dissimilarities between the Ogallala and Permian as exhibited in the contour plots.

The residuals described earlier also can be used to identify unusual patterns. An observed value will be termed unusual if the normalized residual is large, for example, in absolute value greater than two. The term unusual is used in contrast to anomalous because the residuals can be positive or negative. In particular for hydrogeochemical data large negative residuals may to correspond to locations where precipitation from the groundwater is taking place and may be as significant as large positive residuals. The normalized residuals have been coded onto Figure 5 and particularly in the Ogallala, exhibit a pattern which correlates with Areas A and B.

Identifying unusual values by the size of normalized residual is justified for several reasons. In Figure 6, the histograms of the normalized residuals, it is seen that the empirical probability of large residuals is small. If the normalized residuals were distributed normally then the probability of large residuals could be obtained from a normal table. If the residuals are assumed symmetric, then Chebyshev's Inequality asserts that the probability of residuals greater than 2 is less than 0.11. The symmetry that is exhibited in Figure 6 also is indicative of the unbiasedness that should be characteristic of the kriging estimator.

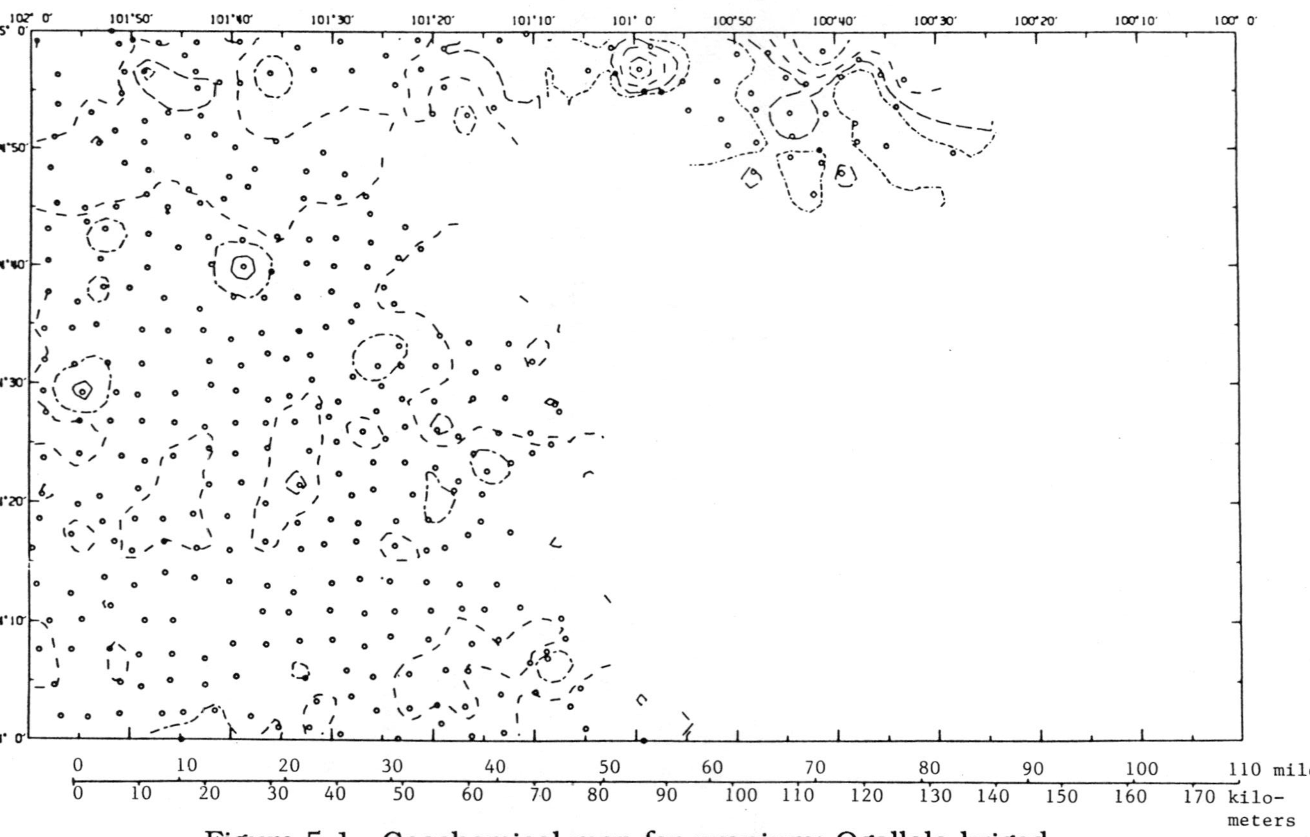

Figure 5-1. Geochemical map for uranium; Ogallala kriged.

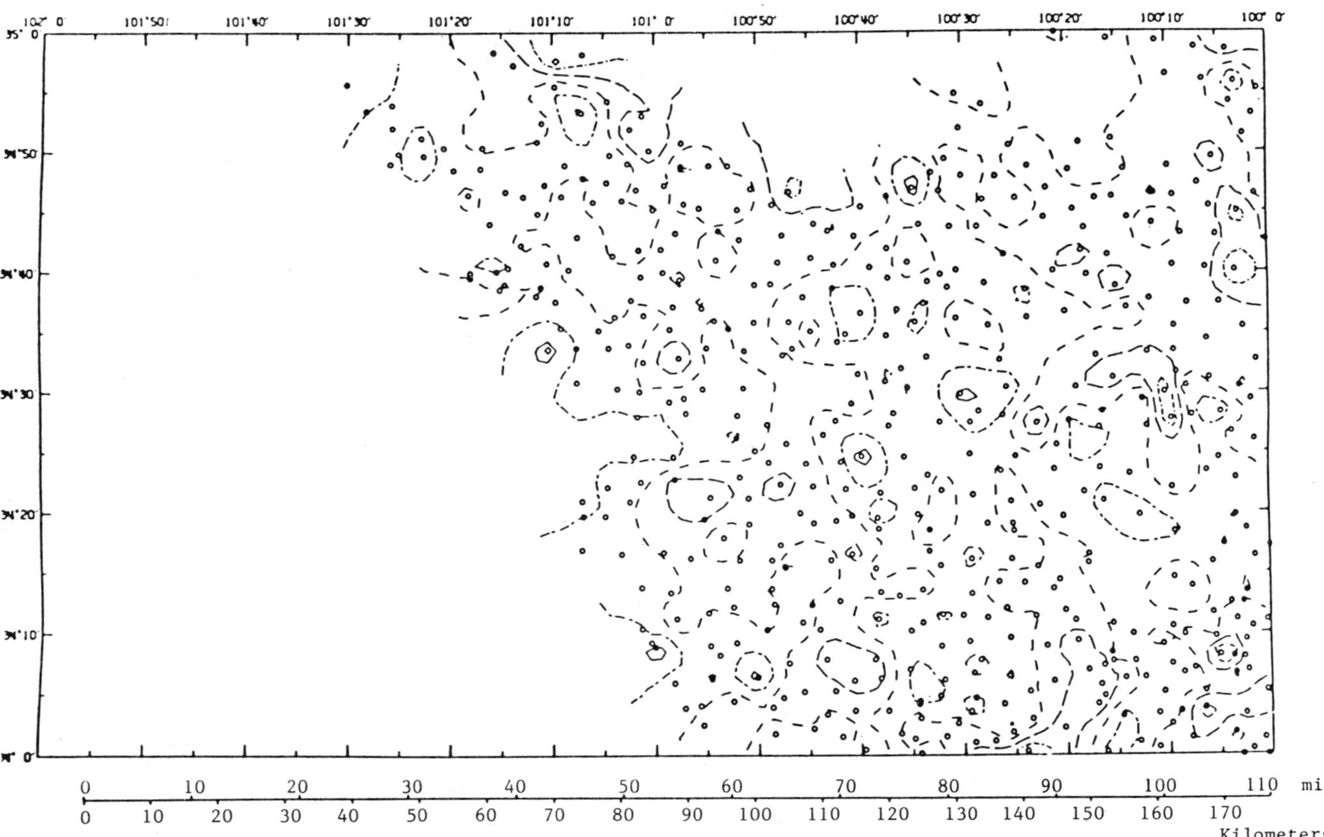

Figure 5-2. Geochemical map for uranium; Permian kriged.

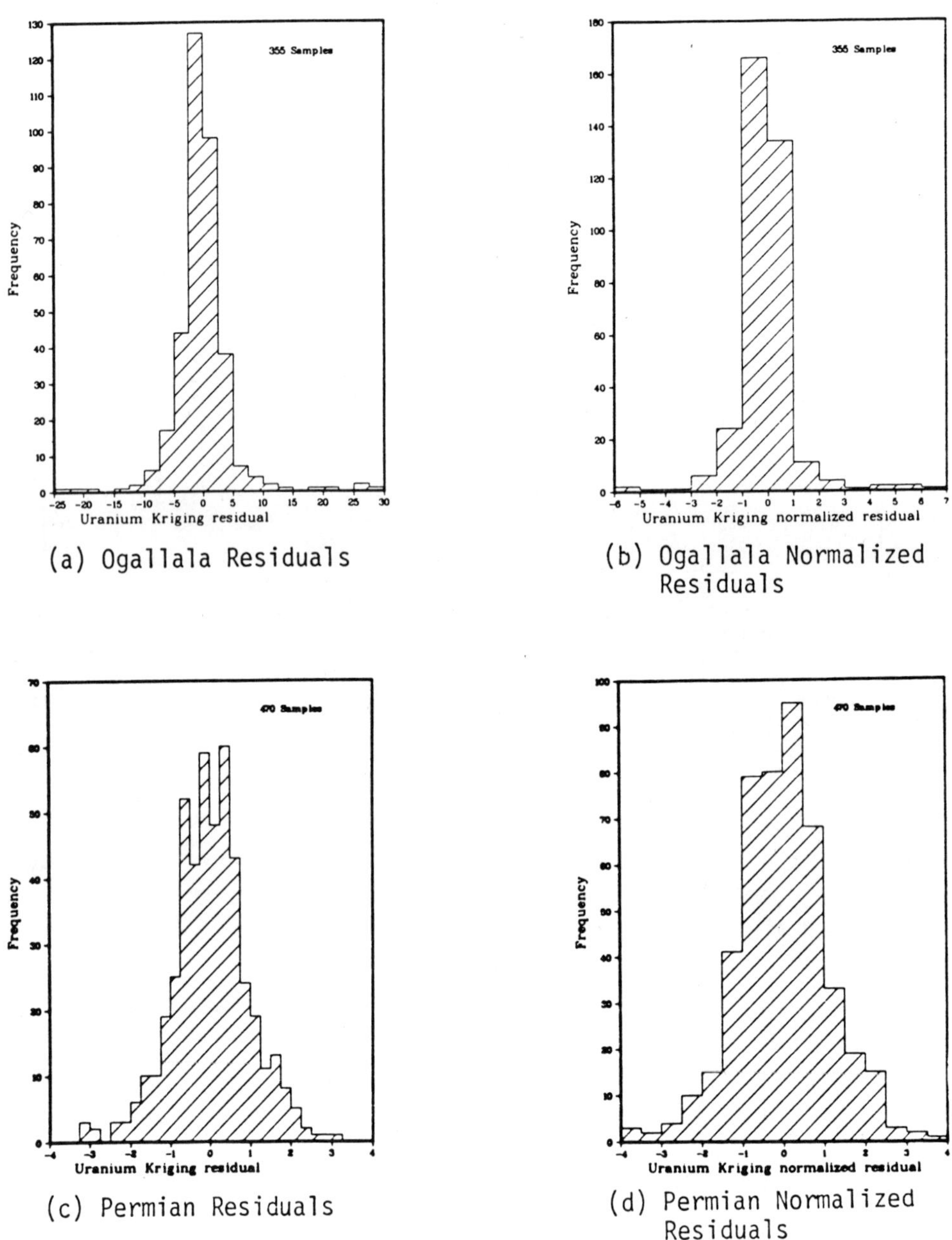

(a) Ogallala Residuals

(b) Ogallala Normalized Residuals

(c) Permian Residuals

(d) Permian Normalized Residuals

Figure 6. Uranium residual histograms from kriging.

The original motivation for studying the variables other than uranium was because of their usefulness in predicting uranium mineralization. However the kriging described does not explicitly incorporate such information. There is a form of joint estimation known as cokriging which was described by Myers (1982, 1983). A somewhat simpler approach was used instead in this preliminary work. Kane (1978) has described the use of weighted sum contouring, using data from the Crystal City and Beeville Quadrangles. When a single variable such as uranium is of principal interest but is known to be related to other variables, weighted sums provide a simple way to incorporate the dependency.

One way to utilize both weighted sums and kriging would be to combine the separate contoured plots but the kriging variances generally would be large. The simpler technique of forming a new variable was used instead, termed Natural Factors. The variables incorporated and the weights are as follows

Permian

1.460* [L-U + (-2.040)] +

2.100* [L-SP + (-8.230)] +

0.610* [L-NA + (-4.540] +

0.610* [L-V + (4.540] +

0.920* [L-MO + (-1.910)]

Ogallala

1.58 • [L-U + (1.80)] +1.30 • [L-LI + (-4.52)] +

1.69 • [L-AJ + (-1.38)] +1.07 • [L-V + (-2.75)] +

1.51 • [L-MO + (-1.98)] + 2.09 • [L-MG + (-3.38)]

In each formation svg's were computed and plotted for the Natural Factors variable. These are shown in Figure 7. A functional model was fitted and used to compute the coefficients in the kriging estimator. Figure 8 shows the composite of the Ogallala/Permian Natural Factors kriged contours overlain with the Amaral favorable areas A, B, C, D and also those identified by Beauchamp and others (1980). Amaral area B is delineated clearly by one or more +4 contours. It is interesting to note that at approximately Lat. 34° 20' N and Long. 191° 10' W there are several +4 contours (Fig. 8). This corresponds to a pattern of large positive normalized residuals and high-level kriged uranium contours as shown in Figure 5. This region does not correspond directly to one of Amaral's areas. It does correspond to Area IIC, IIIC identified by Beauchamp and others (1980).

The Natural Factors contours do not seem to indicate any correspondence with Area D in the Permian. This is not unexpected because the favorable units in Area D are Pennsylvanian in age and are present only at greater depths and are not penetrated by the sampled wells.

Another weighted sum, termed Subjective Mineralization also was kriged but was not as useful.

CONCLUSIONS

Kriging was determined to be a viable geostatistical tool for analyzing geochemical dispersion patterns. The variograms as estimated from the sample variogram plots clearly delineate between variables and geologic units and provide groupings naturally related to predicting Uranium occurrences.

As a tool to identify favorable areas for exploration for uranium several aspects of kriging were utilized; contour plots, normalized residuals, and weighted sum contouring. Coincidence was determined with two areas identified by Amaral (1979) and other areas identified by Beauchamp and others (1980).

Kriging of linear combinations by forming a new variable, as was done with Natural Factors, is not optimal. Neither is kriging of each component. The optimal method is cokriging as is presented in Myers (1982, 1983, 1984) and Carr, Myers, and Glass (1985).

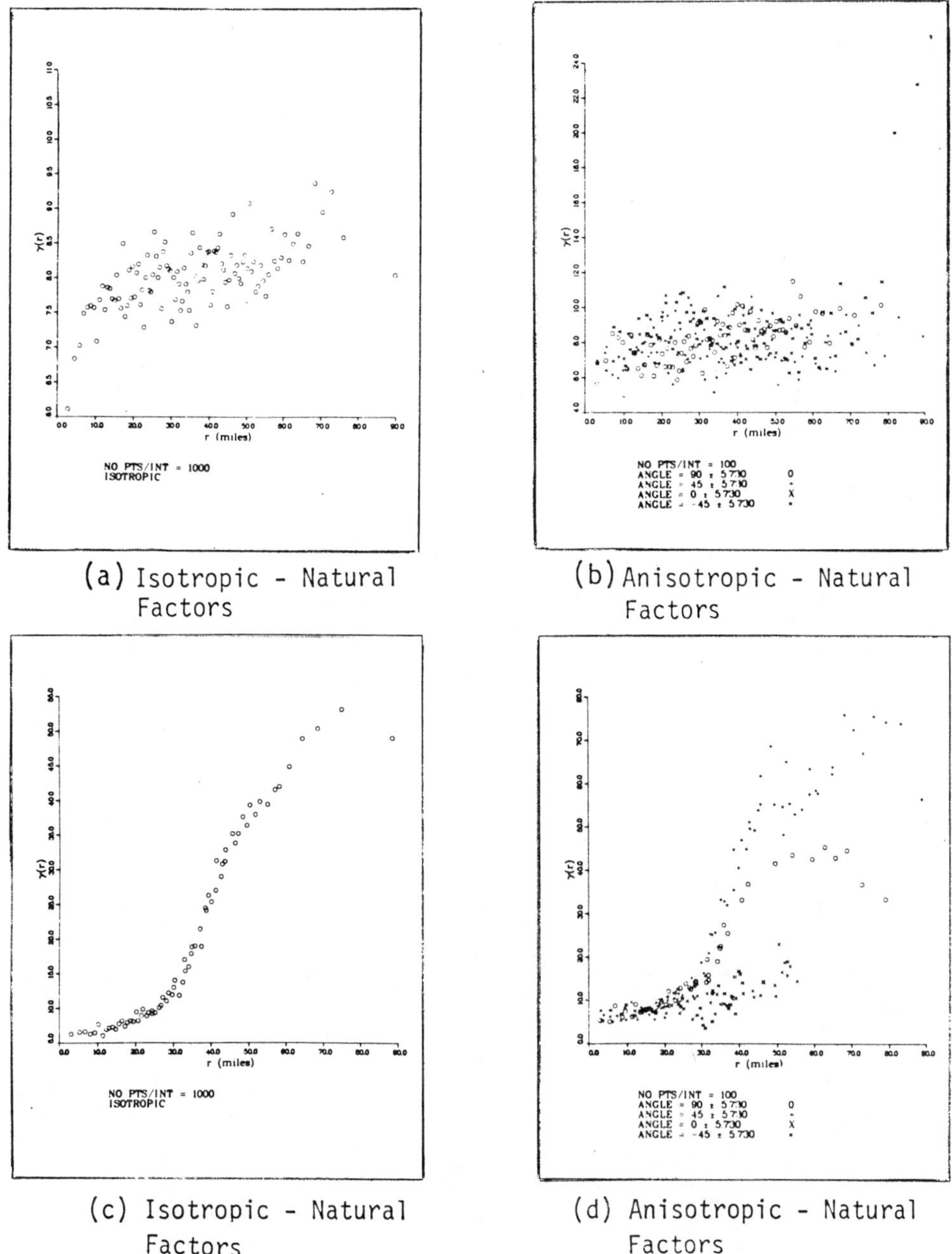

Figure 7. Weighted sum semivariograms for Permian units (a) and (b); weighted sum semivariograms for Ogallala Formation (c) and (d).

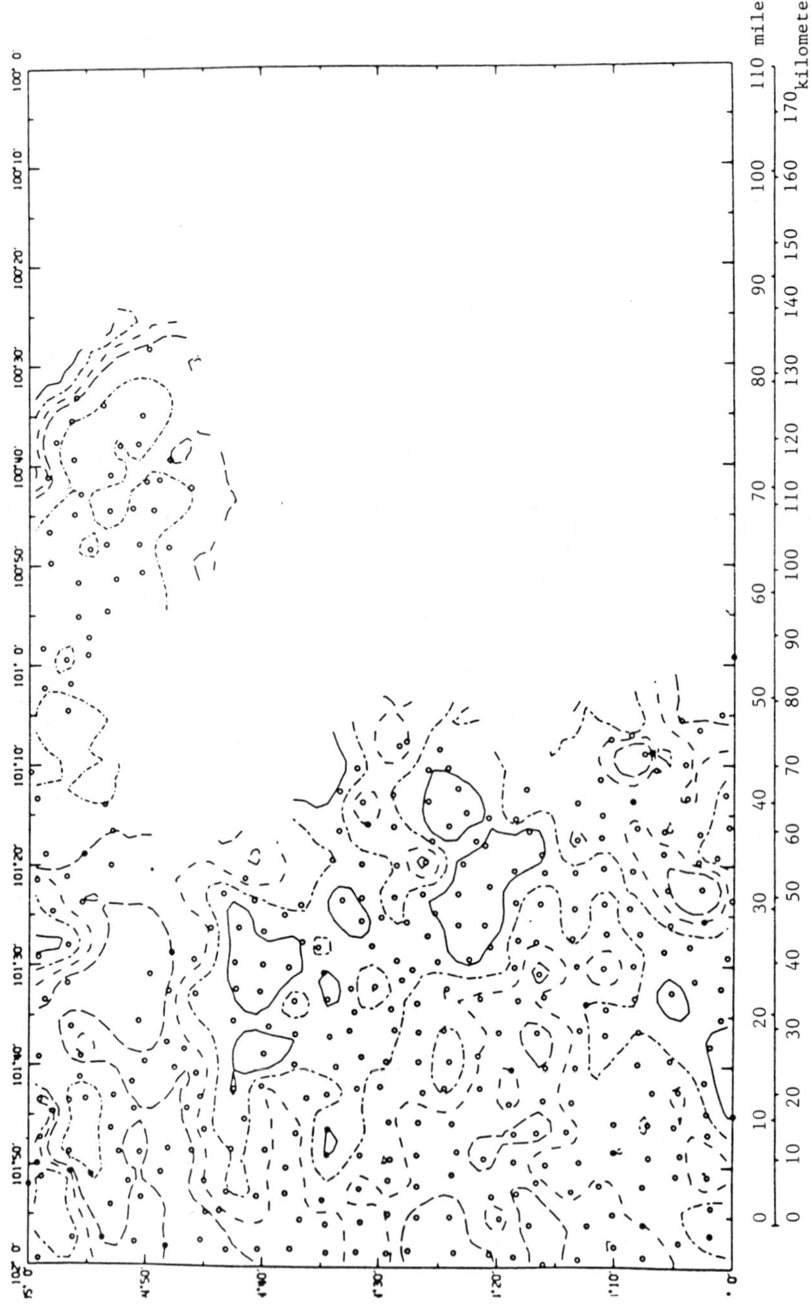

Figure 8-1. Natural factors weighted sum contours from kriging for Ogallala Formation.

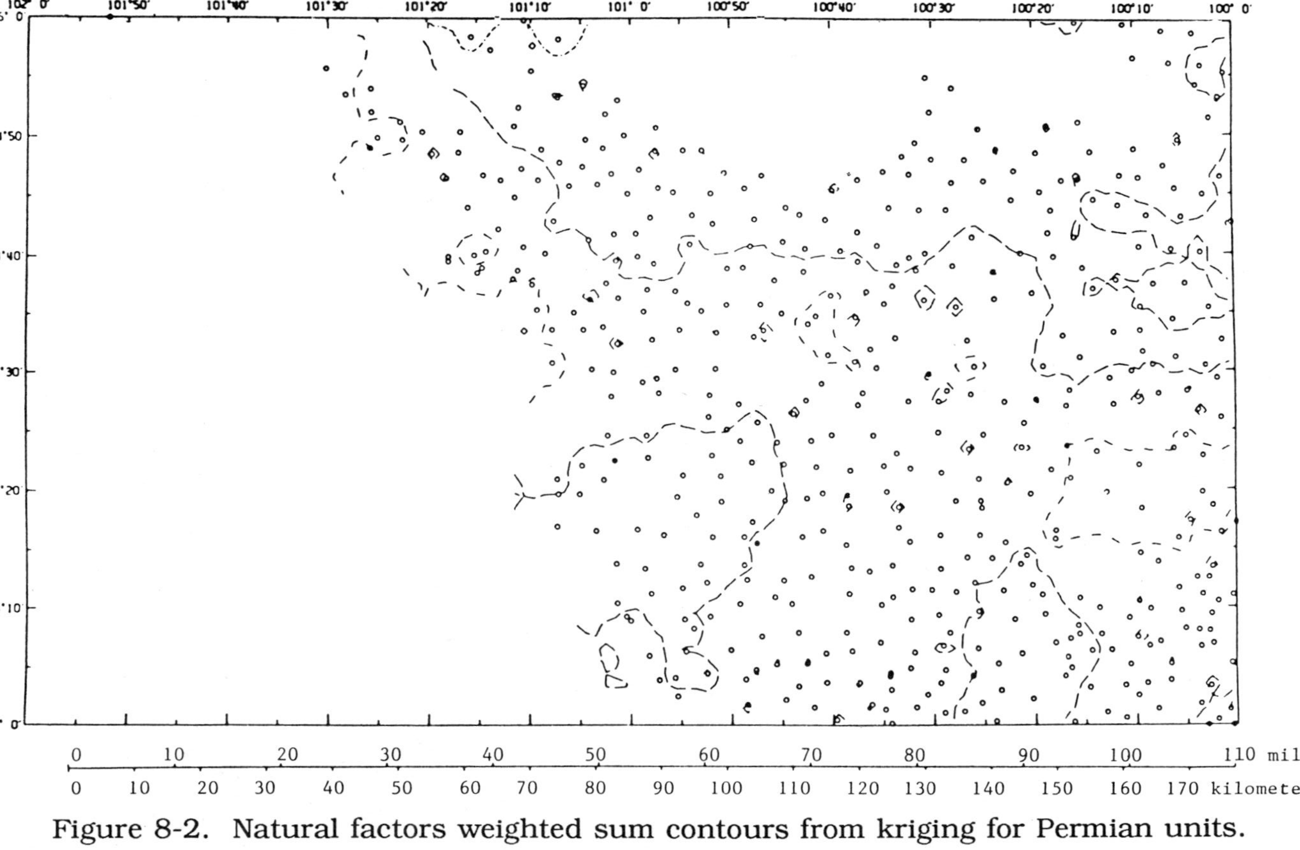

Figure 8-2. Natural factors weighted sum contours from kriging for Permian units.

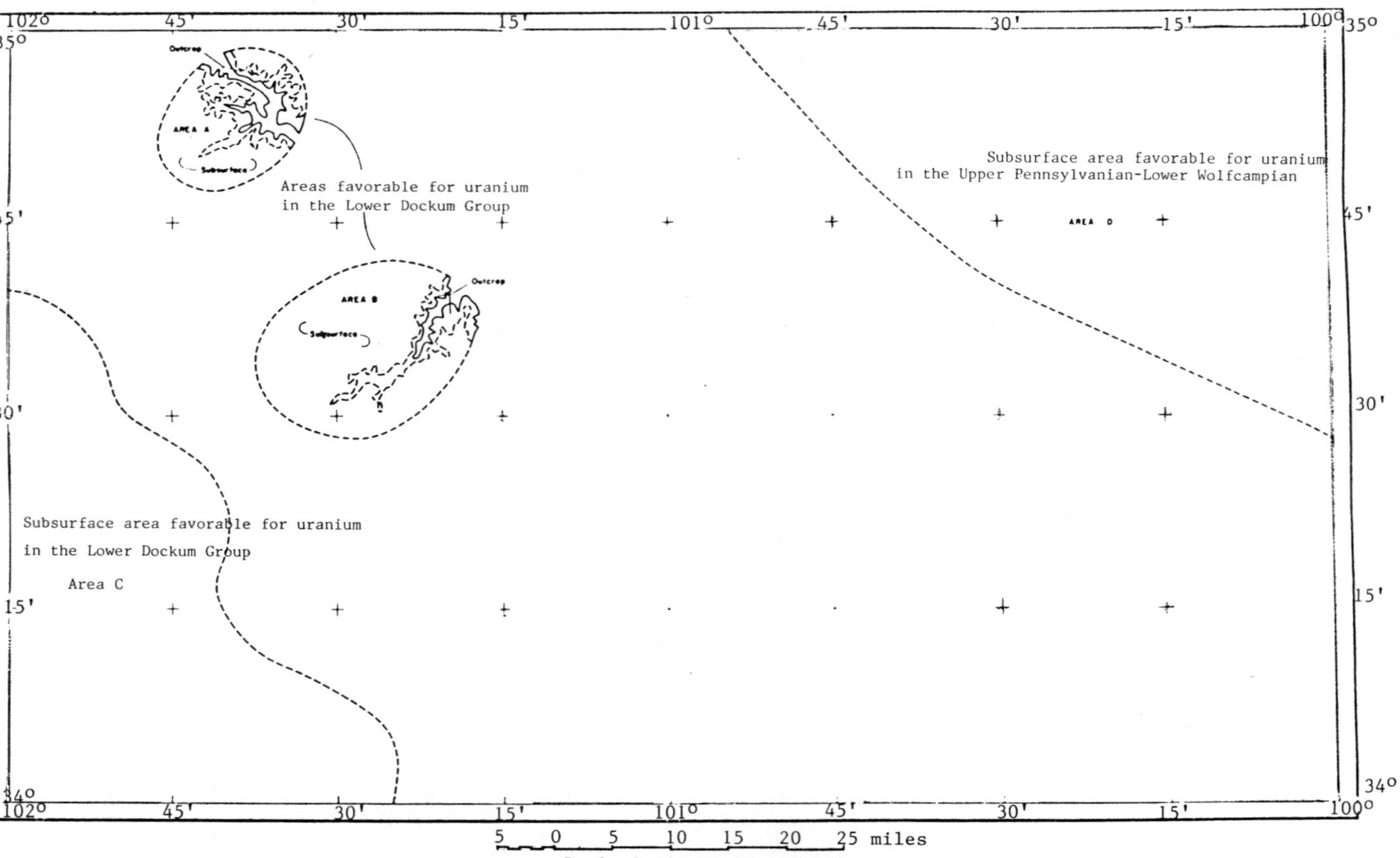

Figure 8-3. Favorable areas of uranium mineralization.

ACKNOWLEDGMENTS

This work was supported by the Oakridge Gaseous Diffusion Plant NURE project.

REFERENCES

Amaral, E. J., National Uranium Resource Evaluation Plainview Quadrangle: Bendix Field Engineering Corp., Grand Junction, Colorado, G. JQ-001(79), 26 p.

Beauchamp, J. J., Begovich, C. L., Kane, V. E., and Wolf, D. A., 1980, Application of discriminant analysis and generalized distance measures to uranium exploration: Jour. Math. Geology, v. 12, no. 6, p. 537-556.

Burgess, T. M., and Webster, R., 1980, Optimal interpolation and isarithmic mapping of soil properties I., The Semi-variogram and punctual kriging: Jour. Soil Science, v. 31, p. 315-331; II., Block kriging: Jour. Soil Science, v. 31, p. 533-541.

Carr J., Myers, D. E. , and Glass, C., 1985, Cokriging - a computer program: Computers & Geosciences, v. 11, no. 2, p. 111-127.

Everhart, D., 1977, Status and progress of the NURE program: Industry Seminar, U. S. Department of Energy, Grand Junction, Colorado, p. 69-102.

Journel, A. G., and Huijbrechts, Ch., 1978, Mining geostatistics: Academic Press, London, 600 p.

Kane, V. E. 1977, Geostatistics: Symposium on hydrogeochemical and stream sediment reconnaissance for uranium in the United States, March 16 and 17, 1977, U. S. Department of Energy, Grand Junction, Colorado. [GJBX-77(77)], p. 203-222.

Kane, V. E., Begovich, C. L. Butz, T. R. and Myers, D. E., 1982, Interpolation of regional geochemistry using optimal interpolation parameters: Computers & Geosciences, v. 8, no. 2, p. 117-136.

Matheron, G., 1965, Les variables regionalisees et leur estimation: Mason et Cie, Paris, 305 p.

Matheron, G., 1971, The theory of regionalized variables and its
    applications: Cahiers du Centre de Morphologie
    Mathematique de Fontainebleau, v. 5, 211 p.

Matheron, G., 1973, The intrinsic random functions and their
    applications: Advances Applied Probability, v. 5, p.437-468.

Myers, D. E., 1982, Matrix formulation of cokriging: Jour. Math.
    Geology, v. 14, no. 3, p. 249-257.

Myers, D. E., 1983 Estimation of linear combinations and co-
    kriging: Jour. Math. Geology, v. 13, no. 5, p. 633-637.

Myers, D. E., 1984, Cokriging-New developments, in Verly, G.,
    and others, eds., Geostatistics for natural resource
    characterization: D. Reidel, Dordrecht, p. 295-305.

Myers, D. E., Begovich, C. L., Butz, T.R., and Kane , V. E., 1980,
    Application of kriging to hydrogeochemical data from the
    National Uranium Resource Evaluation Project: ORGDP,
    Oakridge, Tennessee, K/UR-44, 124 p.

Myers, D. E., Begovich, C. L., Butz, T. R., and Kane, V. E., 1983,
    Variogram models for regional geochemical data: Jour. Math.
    Geology, v. 14, no. 6, p. 629-644.

Roach, C., 1978, Possible NURE resource assessment method-
    ologies: U. S. Department of Energy, Grand Junction,
    Colorado, 11 p.

Uranium Resource Evaluation Project, 1978, Hydrogeochemical
    and stream sediment reconnaissance basic data for
    Plainview NMTS Quadrangle, Texas: ORGDP, K/UR-101,
    36 p.

ANALYSIS OF MASSIVE SULFIDES WITHIN THE MOUNTAIN VIEW AREA OF THE STILLWATER COMPLEX, MONTANA -- A STATISTICAL FORMULATION AND TEST OF THE SULFIDE LIQUID IMMISCIBILITY MODEL

W. J. Bawiec[1], J. H. Schuenemeyer[2], and L. J. Drew[1]

U.S. Geological Survey[1]
University of Delaware[2]

ABSTRACT

The Stillwater Complex, Montana, is a layered mafic intrusion containing resources of chromite, platinum, copper, and nickel. The exploration program for copper and nickel sulfides within the Mountain View area of this igneous body included a series of diamond drillholes located on a grid. Characteristics of massive sulfides which occur in these drillholes (thickness, stratigraphic position, and copper-nickel grades) within and adjacent to the Basal series of the Stillwater Complex have been analyzed and compared with theoretically expected results of immiscible sulfide liquids. An important aspect of this study is the translation of the attributes of a geologic model into statistical hypotheses to evaluate the possibility that liquid immiscibility is the primary process responsible for the thickness, stratigraphic distribution, and copper-nickel grade characteristics of the massive sulfides present within the Mountain View area of the Stillwater Complex.

Geologic models have been proposed on the interrelation between sulfide mineralization and silicate rocks in layered intrusions, mechanisms for separation of immiscible sulfide liquids from basaltic magmas, methods of collection and concentration of immiscible sulfide liquids, and the crystallization of the collected sulfide liquid. The combination of these processes in conjunction with the geologic history of the area leads to expected results concerning the distribution and grade of the massive sulfides contained within layered mafic igneous rocks and the associated metasedimentary rock.

The results of formulating and testing nine statistical hypotheses led to the conclusions that there is no evidence to reject the model wherein sulfide liquid immiscibility is the primary condition responsible for the distribution and grade characteristics of the massive sulfides, and no evidence to reject the hypotheses that the copper and nickel grades of the massive sulfides in the Basal series are not different significantly than those in the metasedimentary rock. Other results of the analysis show that the copper and nickel grades of the massive sulfides are not related to their thickness in either the Basal series or the metasedimentary rock; that no trend in thickness of the massive sulfides was determined as a function of stratigraphic position in either the Basal series or the metasedimentary rock; that frequency in the occurrence of massive sulfides declines as a function of distance from the Basal series and metasedimentary rock contact; and that the total volume of sulfide increases as the depth increases in the Basal series and decreases as distance from the top of the metasedimentary rocks increases.

INTRODUCTION

The application of statistical methods to large arrays of data in evaluating geologic models and processes or describing the distribution of elements within geologic bodies was not possible until the recent advent and availability of computers to the geologist. New avenues of research and investigation concerned with the examination of large data sets have become apparent. This has led to collection of new types of information or a reevaluation of previously collected data, which before, were either too cumbersome to use or too expensive to manipulate. This study is the result of the analysis of detailed geologic information previously collected within the Mountain View area of the Stillwater Complex which contains copper and nickel sulfides. The processes involved are common in many layered magmatic bodies. Therefore, a statistical analysis of the detailed information available from this layered mafic intrusion containing immiscible sulfides could be applicable to other layered intrusions where the data are not as abundant.

Sulfide liquid immiscibility has been suggested by Page (1979) and Zientek (1983) to be an important factor during transportation and emplacement of copper- and nickel-rich sulfides which occur in and adjacent to the Basal series of the Stillwater Complex. This idea is restated and expanded next as the basis for statistical hypotheses used to test and examine the occurrence and grade of copper and nickel sulfides within a set of massive sulfide drillhole intersections from the Mountain View area, Nye, Montana. Massive sulfides are defined as units containing at least 50 percent visual sulfides by volume.

The distribution of copper and nickel concentrations adjacent to and within the Basal series of the Mountain View area of the Stillwater Complex from whole rock analyses of lithologies containing from 0 percent to 100 percent visual sulfides have been described previously (Drew, Bawiec, and Page, 1983; Bawiec and Drew, 1984; Drew and others, 1985; Bawiec, 1985). A correlation has been determined between the magnitude and shape of the copper-nickel concentration patterns and the dominate host lithology, which has led to a correlation between copper-nickel concentration patterns and stratigraphy. A more detailed examination of these concentration patterns as presented in this paper, that is massive sulfides greater than 50 percent visual sulfide, provides a better understanding of the genesis of the ore and the distribution of both copper, nickel, and sulfide.

The liquid immiscibility model is considered as an emplacement condition because of the large volume of sulfide in the Basal series and the adjacent metasedimentary rocks (Page, 1979). Studies suggest that the maximum amount of sulfur soluble in a basaltic magma ranges from 0.05 to 0.2 weight percent as the FeO content ranges from 5 to 20 weight percent (Haughton, Roeder, and Skinner, 1974) at 1,200 degrees centigrade. Other studies suggest that basaltic magma becomes saturated with sulfide at 0.038 weight percent and 1065 degrees centigrade (Skinner and Peck, 1969), which is not sufficient to account for all the sulfide present in the Stillwater Complex. Also, sulfur-isotope analyses and sulfide-mineral compositions indicate that assimilation of sulfur and metallic elements from underlying country rocks is not responsible for the basal-sulfide concentration (Page and others, 1985). Therefore, it is concluded that the sulfides were derived from a sulfide-saturated magma containing immiscible sulfide liquid droplets, which are indicated by the overall distribution of the sulfide and the occurrence of small inclusions and clusters of polymineralic sulfides within the centers of cumulus plagioclase and pyroxene grains (Page, 1979).

This mixture of sulfide-saturated basaltic magma and immiscible sulfide liquid is assumed to have been injected into the metasedimentary rocks. Within the main chamber, local crystallization and accumulation of cumulates began. Movement within the magma resulted in collision and coalescence of sulfide-oxide droplets influenced more and more by gravity as they grew larger. This resulted in a gravitational accumulation of the sulfides that migrated toward the base of the igneous rocks (Page, 1979).

A variation on the emplacement history as discussed previously by Page (1979), was introduced by Zientek (1983). Zientek agreed that sulfides were derived from a sulfide-saturated magma containing immiscible sulfide liquid droplets and that sulfide was present immediately after emplacement. However, Zientek proposed that the occurrence of sulfides in the metasedimentary rock is due mainly to the emplacement of an initially sulfide-enriched magma, which resulted in the formation of sulfide-rich sills and dikes in the metasedimentary rocks that crystallized to form mafic norites and adjacent massive sulfides. As magma continued to be injected to form the Basal series, the proportion of sulfides in the magma became less.

The purposes of this paper are to examine the distribution, associated thicknesses, and copper and nickel grades of massive sulfides in the Mountain View area and to compare these attributes with expected results of immiscible magmatic sulfides. Because many fluctuating variables, silicate/sulfide ratios, partitioning coefficients, influxes of new magma, changes in temperature, pressure, and composition, etc., have a significant effect on the resulting distribution of both the massive sulfides and the copper-nickel grades contained within them, it is difficult to evaluate one primary process and disregard all others. However, given a specific set of conditions that probably operated within a specific area, expected results of a process can be evaluated.

GEOLOGIC SETTING AND STRATIGRAPHY

The Stillwater Complex, a large tabular mass of layered Precambrian mafic and ultramafic rocks, is exposed along the

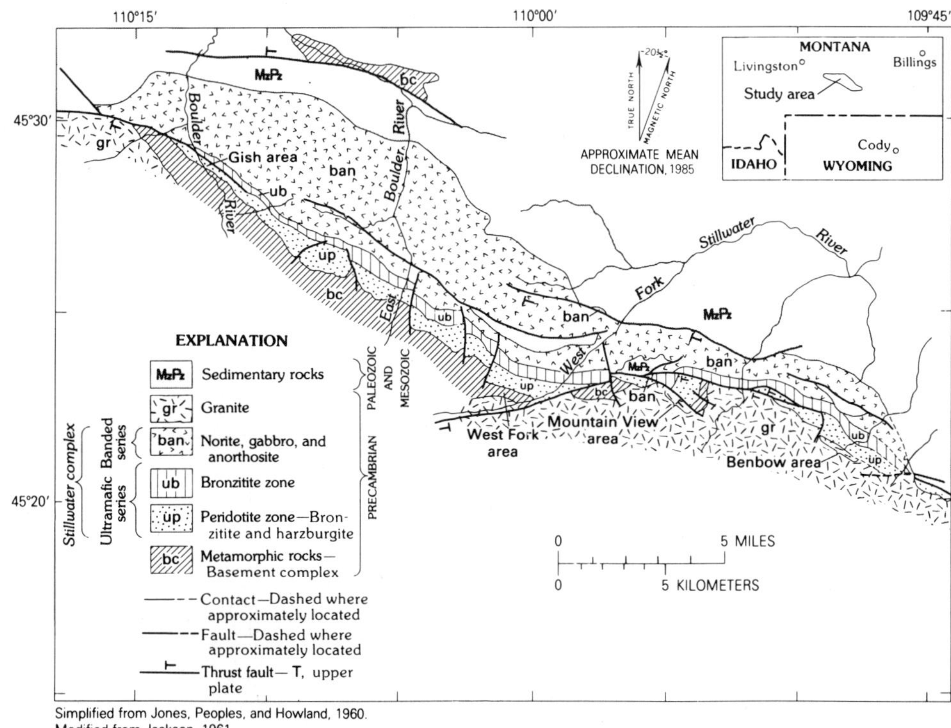

Simplified from Jones, Peoples, and Howland, 1960.
Modified from Jackson, 1961.

Figure 1. Geologic index map of Stillwater Complex, Montana,
showing location of Mountain View area.

northern border of the Beartooth Mountains in the southwestern
corner of Montana (Fig. 1). The exposure is approximately 48 km
long and has a maximum exposed thickness of 5.5 km. The
layering in the complex strikes northwest and dips northeast. As
the Stillwater Complex is exposed only partly and its upper units
are eroded, its original size and shape are unknown.

Various terminologies have been devised to describe the
stratigraphic succession in the Stillwater Complex (Jones,
Peoples, and Howland, 1960; Jackson, 1961; Hess, 1960; Page,
1977, 1979; McCallum, Raedeke, and Mathez, 1980; Segerstrom
and Carlson, 1982; and Todd and others, 1982), but we have
elected to follow the terminology used by Zientek, Czamanske,
and Irvine (1985). The differentiated, stratiform complex is
divided into five main parts: the Basal, Ultramafic, and Lower,
Middle, and Upper Banded series (Fig. 2). This report is
concerned with the Basal series and the underlying
metasedimentary rocks. The Banded series and the Ultramafic

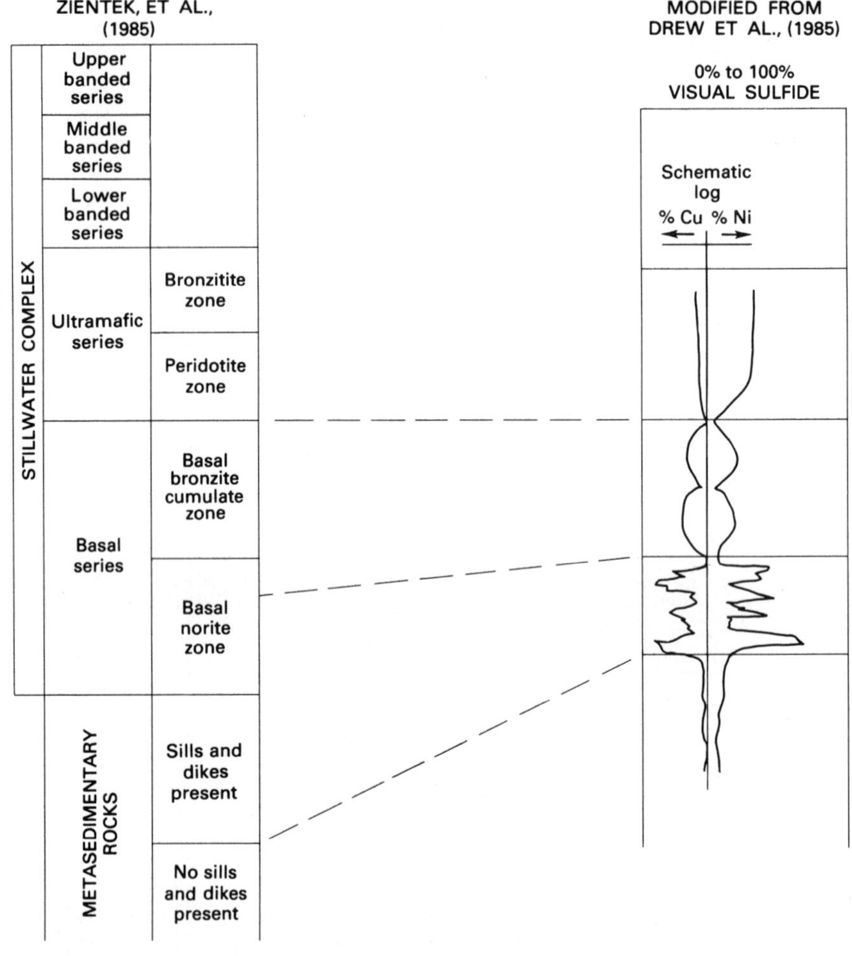

Figure 2. Generalized stratigraphy of Stillwater Complex and characteristic copper and nickel concentration patterns.

series are higher stratigraphically in the section and were not included in this study.

The Basal series has been divided into two zones, with the Basal bronzite cumulate zone lying above the Basal norite zone (Zientek, Czamanske, and Irvine, 1985). The Basal series in the Mountain

View area consists dominantly of orthopyroxene-rich cumulates
containing xenoliths of metamorphosed sedimentary rocks. The
Basal norite zone is composed of alternating lensoid masses or
layers of rock containing olivine, orthopyroxene, clinopyroxene,
plagioclase, hornblende, biotite, quartz, iron-titanium-chromium
oxides, and sulfide minerals (Page, 1979) in which a variety of
crystallization sequences occur. Cryptic variation of
orthopyroxene compositions and the localized variety of
crystallization sequences indicates magma mixing, multiple
injections, and slight local differences in magma composition.

The Basal series is underlain by the metasedimentary rock that it
intrudes (Fig. 2). Most of the exposed metasedimentary rock
below the base of the Stillwater Complex consists of fine- to
medium-grained unlayered or massively layered rocks, but
compositional layering is well developed in our area of study, the
Mountain View area (Page, 1977). These layered
metasedimentary rocks were folded into major and minor open
isoclinal folds where axis trend northeast and plunge 60 degrees
NE. These initial folds were warped or refolded into broad open
folds about axes that now plunge moderately northwest (Page,
1977).

The sills and dikes associated with the Stillwater Complex (Fig. 2
and 3) are restricted to a zone in the metamorphosed
sedimentary rocks, generally less than 200 meters wide, that are
adjacent to the basal contact of the Stillwater Complex (Zientek,
1983). On the basis of field criteria, two distinct types of sills and
dikes can be recognized. Diabases are sulfide-poor,
diabasic-textured gabbro norites and norites that show uniform
mode, grain size, and texture. Mafic norites are sulfide-bearing
orthopyroxenites, melanorites, and norites that show variable
modes, textures, and grain size within and between intrusive units
(Zientek, 1983).

The Mountain View area is a triangular-shaped area (Fig. 1)
bounded by two high-angle reverse faults; the Lake Fault and the
Bluebird Thrust (Fig. 4). As a consequence of rotation of this
fault-bounded block, the stratigraphy strikes northeast and dips
to the northwest. The rectangle in Figure 4 shows the perimeter
of the drillhole grid within the location of the copper and nickel
exploration target of the Mountain View area. This rectangular
area is defined by a grid upon which over 108 diamond drillholes
are located. The stratigraphy and fault intersections identified
within 87 vertical drillholes (used in conjunction with and

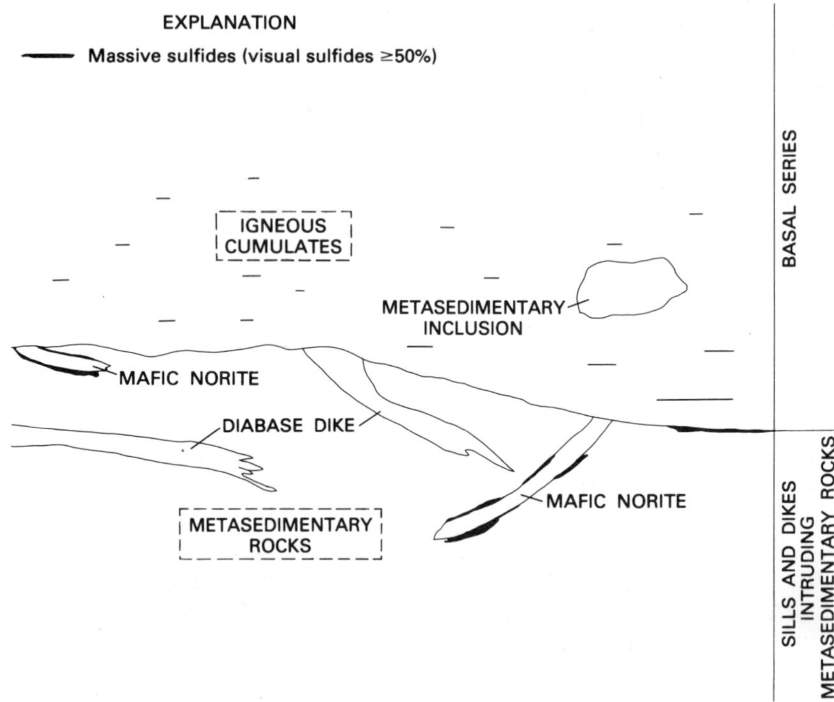

Figure 3. Idealized section showing expected distribution of massive sulfides from the sulfide liquid immiscibility model.

constrained by the previously mapped surficial geology) resulted in a three-dimensional structural interpretation of the Mountain View study area.

DATA SOURCE AND ACKNOWLEDGMENTS

The drillhole data used for this study are from the Mountain View area in Stillwater Country, Montana. The data set contains information from 108 diamond drillholes on a grid in which a hole spacing is approximately 200 feet and depths are a maximum 1,700 feet. The information for each drillhole consists of lithologic and structural descriptions plus copper and nickel whole-rock concentration values for each interval sampled. This

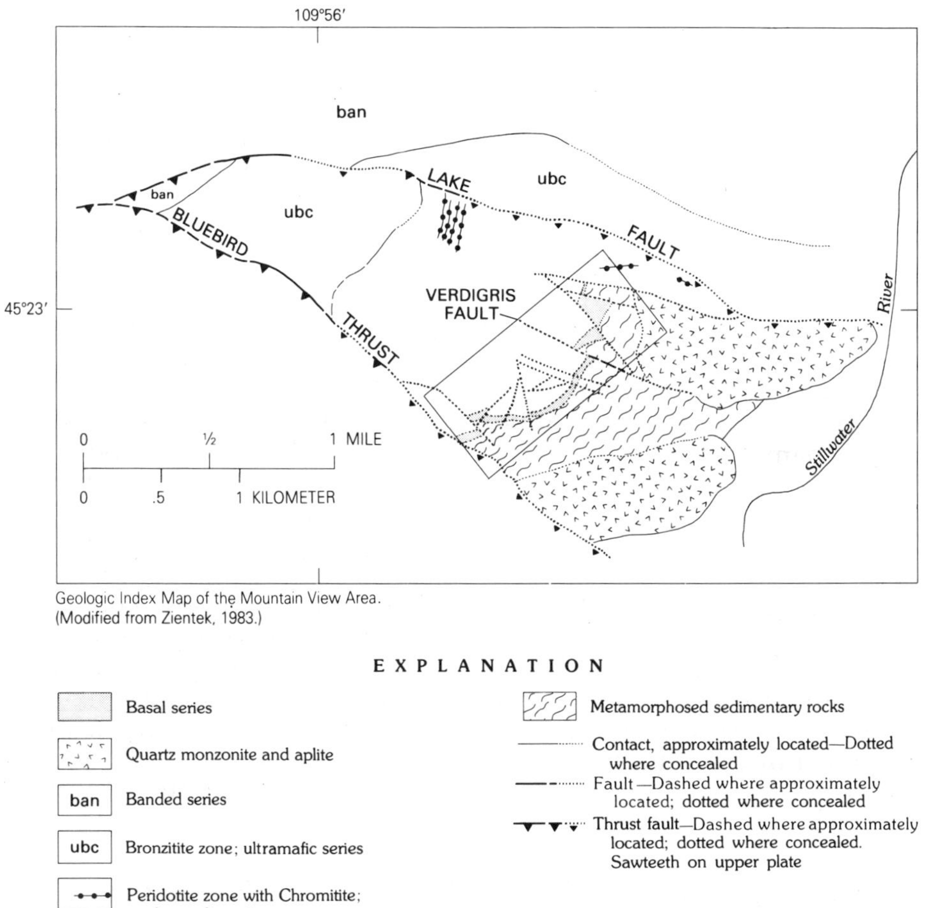

Geologic Index Map of the Mountain View Area.
(Modified from Zientek, 1983.)

EXPLANATION

Basal series

Quartz monzonite and aplite

ban   Banded series

ubc   Bronzitite zone; ultramafic series

Peridotite zone with Chromitite; ultramafic series

Metamorphosed sedimentary rocks

Contact, approximately located—Dotted where concealed

Fault—Dashed where approximately located; dotted where concealed

Thrust fault—Dashed where approximately located; dotted where concealed. Sawteeth on upper plate

Figure 4. Geologic index map of Mountain View area, Stillwater Complex, Montana, showing location of study area.

information was collected during a long period of time and detailed investigations were conducted by many people, especially Norman Page, Michael Zientek, and Roger Cooper. The authors are grateful to Anaconda Minerals Company, especially Roger Cooper and Alistair Turner for access to and use of these data, and to Michael Zientek and Norman Page for critical evaluations and helpful suggestions in reviewing this paper.

DATA PREPARATION

Owing to the volume of lithologic, structural, and copper-nickel concentration data made available for this study, the analysis of these data required computerization. The primary data sets for each drillhole consisted of structural, lithological, and copper-nickel-concentration files, which were combined into a composite log with the resolution linked to the length of the interval sampled. Of the 108 diamond drillholes used in this study, the average length of core analyzed for copper and nickel concentrations was approximately 5.5 feet. Also, no detectable pattern to variations from this 5.5 foot average was located; that is sampling intervals did not become smaller near the massive sulfides. To evaluate the sulfide liquid immiscibility model, we examined data on the contact between the igneous lithologies (Basal series) and the metasedimentary lithologies below the base of the complex intruded by associated sills and dikes. Examination of 108 drillholes showed that 34 holes penetrated the lower boundary of the Basal series. These 34 holes were examined; any contacts that were the result of faulting were eliminated in order to discard lithologies out of stratigraphic position and at distorted distances from the contact. The location of faults above and below the Basal series/metasedimentary rock contact encountered in each hole also was noted in order to set limits on undisturbed lithologies to be investigated. The Basal series/metasedimentary rock contact appeared unfaulted in 28 drillholes.

Using this subset of 28 drillholes containing unfaulted lower contacts of the Basal series, we then investigated the distribution and occurrence of sulfides. Massive sulfides were defined as intervals containing minimum visual sulfide of 50 percent. Selections of intervals containing 50 percent or more visual sulfides resulted in 18 observations in the Basal series and 25 observations in the metamorphosed sedimentary rocks intruded by sills and dikes, all distributed among 20 drillholes (Fig. 5). hence, 8 drillholes did not contain intervals of visual sulfide greater than or equal to 50 percent. Because of the resulting small number of observations, statistical tests have minimal power; however, significant trends were observed.

The selection of a cutoff of 50 percent visual sulfide was based upon the following considerations. First, we wanted to analyze only one population, massive sulfides. If we lowered the amount

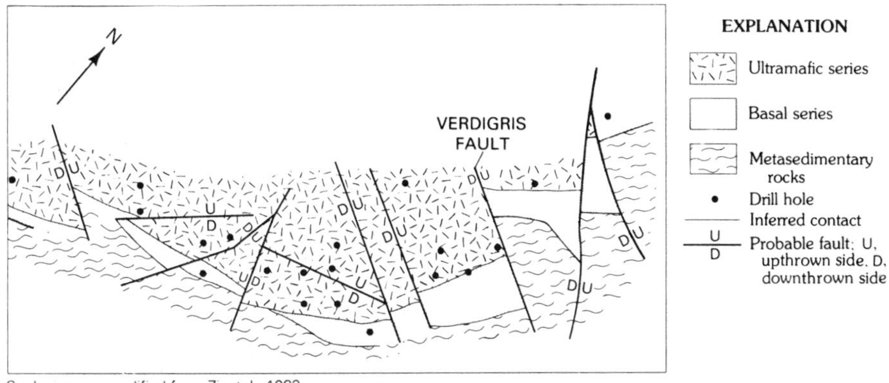

Geologic map modified from Zientek, 1983.

Figure 5. Geologic index map of study area, Mountain View area, Stillwater Complex, Montana, showing locations of drillholes and fault-bounded blocks.

of visual sulfide below 50 percent by volume we might include units other than those of our main interest. Sulfides associated with the sills and dikes, in most places, are either present and constitute a high proportion of the unit, or absent. However, in the Basal series the gradational change ranges from 0 to 100 percent. Therefore, if the visual sulfide minimum is lowered, disproportionally thick intervals from the Basal series would be included.

In order to examine the true thicknesses and distances of these sulfide-rich intervals from the Basal series/metasedimentary rock contact, we had to consider the present orientations of the rocks and the attitude of the drillholes. This was done by examining and interpreting the structure of the study area, computing the orientations of each inclined fault-bounded block of ground, and adjusting the apparent thicknesses and distances to true thicknesses and distances.

The grades for copper and nickel within each of these true-thickness massive sulfide intervals are computed by taking a weighted average of the copper-nickel concentration of all the samples occurring within the intervals. Because the massive sulfide intervals range from 50 percent to 100 percent visual sulfides, it was expected that the higher grades would be associated with intervals containing the higher proportion of sulfides. However, no significant correlation could be determined between grades for copper and nickel, and the proportion of sulfides which occur within the massive sulfides.

THE MODEL

The sulfide liquid immiscibility model and expected results are a consequence or product of three conditions of emplacement: the geologic history of the area; the initial mode or distribution of the sulfides; and the crystallization sequence for immiscible magmatic sulfides. Admittedly, additional geologic evidence that supports, refutes, or complicates the model as proposed here has been omitted because such evidence lies outside the scope of this study, which is a description of the distribution of massive sulfides, related copper-nickel grades, and demonstration of the statistical formulation and testing of geologic hypothesis.

Significant events within the geologic history of the Mountain View area include: (3140 m.y. ago) deposition of sediments from a mafic or ultramafic source into one or multiple local basins; low-grade metamorphism and isoclinal folding of rocks; (2700 m.y. ago) intrusion by Stillwater magma that produced a metamorphic aureole in the folded sedimentary rocks followed by intrusion of quartz monzonite plutons; deposition of marine and continental sedimentary rocks; and Laramide deformation (Page, 1977, 1979).

The influence of these events on our model relates mainly to the distribution of massive sulfides within the metasedimentary rock. No known control has been identified to influence either the distribution or the thickness of sulfides in the metasedimentary rock; however, an association has been determined between the occurrence of mafic norite sills and dikes and the massive sulfides adjacent to them (Zientek, 1983). A relationship also has been determined between the occurrence of sills and dikes in the metasedimentary rock and their distance from the base of the Basal series. Sills and dikes occur no farther than 200 meters below the base of the Basal series (Zientek, 1983). For this reason, a greater number of massive sulfides are expected to be proximal to the base of the Basal series than farther away. Also, the mixing of sulfides during transportation and emplacement, and the lack of vertical control on the emplacement of massive sulfides within the metasedimentary rocks would result in a homogenization of the copper and nickel grades and elimination of any trends in the distribution of grades of the massive sulfides.

The mode or distribution of immiscible sulfide liquid contained within the silicate magma can have two dispositions: (1) a homogeneous mixture of immiscible sulfide droplets and sulfur-saturated silica magma could be injected into the metasedimentary rocks and the magma chamber in multiple pulses, where collision and coalescence of the sulfide droplets would cause their gravitational migration toward the base of the complex (Page, 1979), or (2) an intrusion of sulfur saturated silica magma containing an initially larger proportion of immiscible sulfide liquid could be injected into the metasedimentary rock and create sills, dikes, and adjacent massive sulfides, with later pulses of silicate magma containing proportionally less immiscible sulfide liquid (Zientek, 1983).

A consequence of the distribution of massive sulfides, in either disposition, would be the same: a greater volume of sulfides occurring lower in the igneous rocks. However, Page's explanation would require more vertical movement of sulfides because Page assumed initial homogeneous distribution of sulfides and gravitational migration within the magma, whereas Zientek's explanation would require a variable supply for the source of the sulfides with immiscible sulfides that are greater proportionally during initial intrusion. Both of these possible dispositions, accompanied by vertical movement which is restricted, could lead to variation in accumulating massive sulfide thicknesses. A by-product of the presence of a large amount of sulfide in the system could be dilution of grade due to the limited amount of copper and nickel available to be incorporated into the sulfide. This also would result in an apparent reduction in the amplitude of trends which would be present if less sulfide were present.

The crystallization sequence for immiscible magmatic sulfides and the influence this has upon copper and nickel grades is an intricate process. The depositional patterns of many ores are such that even a single hand specimen may represent a complex series of superposed chemical systems (Barton, 1970). However, many of these complexities are concerned with re-equilibration of sulfides as temperature drops and are of a scale too detailed to concern this study. We are concerned with gross mineralogical trends, which are detectable in drillhole data where a sampling interval averages 5.5 vertical feet. These trends involve the distribution of the occurrence of massive sulfides and their related copper and nickel grades.

The expected order of events in the crystallization sequence of magmatic sulfide within the copper-iron-nickel system as stated here has been described by Craig and Kullerud (1969). As a result of fractional crystallization, formation of a liquid relatively rich in nickel may be possible above 1000 degrees centigrade with the crystallization of hexagonal pyrrhotite. As the temperature drops to 850 degrees centigrade, separation of liquid that is richer in copper than is the crystallizing pyrrhotite may occur because of the increasing amount of nickel contained within the pyrrhotite. This segregation of a copper-rich liquid may result in formation of chalcopyrite or copper-rich ores (Hawley, 1962). Below this temperature as the immiscible sulfides continue to crystallize, monosulfide solid solution (MSS) forms, possibly as a homogeneous phase. As the temperature drops, a low-temperature re-equilibration of the MSS into monoclinic pyrrhotite and chalcopyrite-pentlandite mineral assemblage alters the mineral compositions subsequent to their original emplacement. This can occur through exsolution of pentlandite from pyrrhotite, and hence results in the reduction of nickel content in the pyrrhotite.

The products of this crystallization sequence of immiscible magmatic sulfides can be diverse. Characteristics of accumulating massive sulfides can be affected by processes operating throughout the entire crystallization sequence. For example, ideally the richest grade sulfides would be expected at the base of the igneous rocks because they probably would have access to the most available metals, undergo the longest magma-sulfide interaction time, and travel both short and long distances through the magma chamber. This expected distribution of copper and nickel grades may be modified, however, because the vertical flow could be inhibited by new pulses of magma, the possible scouring of pooling or pooled sulfides could lead to resuspension, and sulfides could be remobilized due to filter pressing of cumulus piles. These processes also could affect characteristics related to thickness of the massive sulfide, distance from the base of the igneous rocks, and copper-nickel grades. However, even with these limitations, an investigation and analysis of the massive sulfides contained within the igneous and metasedimentary rocks of the Mountain View area is valuable, even if only to document spatial characteristics such as occurrence, thickness, and grade.

Possible explanations to describe how these distributions occurred may be subject to change and reinterpretation, but the quantitative characteristics of the massive sulfides such as distribution, thickness and copper and nickel grades will not change.

THE HYPOTHESES

Nine hypotheses describing the expected occurrence and grade characteristics of the massive sulfide bodies were divided into four related groups. These hypotheses were constructed and tested from this model in which the expected spatial distribution of massive the sulfides is as shown in Figure 3. The hypotheses that are inferred from the liquid immiscibility model are denoted $H_0(i)$, i = 1, 2, ..., 9. The alternative hypotheses are the negation of these null hypotheses. Some null hypotheses suggested by the liquid immiscibility model were exchanged with the alternative to perform the statistical tests. For example, the model suggests that the copper and nickel grades of the massive sulfides in the Basal series are correlated. In the statistical test the null hypothesis is zero correlation.

The first group consists of three hypotheses that evaluate the copper and nickel grades in the massive sulfides as a function of their stratigraphic position. These hypotheses were constructed to test whether the copper and nickel grades in the massive sulfides of the igneous Basal series and metasedimentary rocks were similar. If the sulfides have the same genesis or result from the same magma pool with no subsequent alteration, then the characteristics of copper and nickel grades should be the same. If the sulfides of the metasedimentary rock came from a different sulfide source, they could show different copper and nickel grades. It was hypothesized that the sulfides in the Basal series and in the metasedimentary rocks are the same; therefore, the copper and nickel grades of the massive sulfides in the Basal series do not differ significantly from the copper and nickel grades of the massive sulfides in the metasedimentary rocks; this will be referred to here as hypothesis $H_0(1)$. Second, it was hypothesized from the liquid immiscibility model that the copper and nickel grades are not related to their distance above the base of the Basal series, $H_0(2)$. Although zoning of copper, nickel, and copper/nickel ratios has been described for Strathcona (Cowan, 1968), Noril'sk-Talnakh (Genkin, Eustighoeva, and Kovalenker, 1980), and Frood (Hawley, 1965) it has not been observed in many other deposits which implies it requires special conditions of brecciation and fracturing of the footwall if the fractionating sulfide liquid is to be filter pressed away from early crystallizing monosulfide solid solution (Naldrett, 1981). Similarly, it is hypothesized that this same lack of trend in grades would be observed in the massive sulfides associated with sills and dikes of the metasedimentary rocks as a function of their distance from the top of this unit, $H_O(3)$ because of the lack of control on the vertical distribution or emplacement of the massive sulfides.

The second group consists of two hypotheses concerning the relationships between copper and nickel grades and massive sulfide thicknesses. From the described model, it was postulated that there would be no significant correlation between copper grade and thickness and nickel grade and thickness of the massive sulfides in the Basal series, $H_O(4)$, and the massive sulfides in the metasedimentary rocks, $H_O(5)$. This would be due primarily to the large sulfide volume, multiple injection of new magmas, possible convection processes differentially inhibiting and accelerating sulfide movement, remobilization and concentration of sulfides due to filter pressing of cumulate piles in the Basal series and the lack of control on vertical distribution, and thickness of massive sulfides in the metasedimentary rock.

The third group consists of two hypotheses concerning trends in the volume of the massive sulfides as a function of distance from the contact between the base of Basal series and the metasedimentary rocks intruded by sills and dikes. First, it was hypothesized that the thickest concentrations of massive sulfides in the Basal series would occur close to this contact because it probably represents a surface upon which the sulfide magma tended to collect. Thus, massive sulfides in the Basal series probably would become thicker and more frequent; as a result, greater volumes of sulfides would be closer to the base of the Basal series, $H_O(6)$. Second, it was hypothesized that the occurrence of sulfide magma in the metasedimentary rocks would become less as distance increases below this same contact; therefore, occurrence would be less frequent and the volume of sulfides would decrease as distance from this contact increases, $H_O(7)$. Massive sulfides in the metasedimentary rock are associated with the sills and dikes. The sills and dikes, in turn, are associated spatially with the base of the Basal series and occur within 200 meters (Zientek, 1983). Although unproven, here it is assumed that the occurrence of sills and dikes become less abundant as distance from the base of the Basal series increases.

The fourth group consists of two hypotheses which test whether the copper grade is correlated to the nickel grade in the massive sulfides of the Basal series, $H_0(8)$ and if the copper grade is correlated with the nickel grade in the massive sulfides associated with sills and dikes, $H_0(9)$. These correlations would be a consequence of the manner by which both copper and nickel would be partitioned into the sulfide phase during crystallization.

TEST OF HYPOTHESES

The nine hypotheses are addressed individually by stating the null hypothesis, the reason for the expected results, the test of hypothesis, and the conclusion.

> (1) $H_0(1)$, the copper grades of the massive sulfides in the Basal series are not different significantly than the copper grades of the massive sulfides associated with the sills and dikes of the metasedimentary rocks; the nickel grades of the massive sulfides in the Basal series are not different significantly than the nickel grades of the massive sulfides associated with the sills and dikes of the metasedimentary rocks.

This result is expected if the immiscible sulfides in both units came from the same sulfide source without further alteration.

A visual indicator of the test of this hypothesis was shown by constructing and comparing box plots of the copper and nickel grades of massive sulfides within the Basal series and the associated sills and dikes (Figs. 6A and 6B). In these figures, the sides of the box (i) represent the lower and upper quartile, the plus sign (+) represents the median, and the (*) and o's suggest possible outliers. In addition, box plots were constructed and compared for percentages of visual sulfides (50 percent to 100 percent) and thickness of the massive sulfides (Fig. 6C and 6D). Comparison of these four pairs of box plots show the median in the sills and dikes to be shifted to the right and, therefore, having values higher than those in the Basal series. However, these displacements are so small that there is a high probability that this could have occurred by chance.

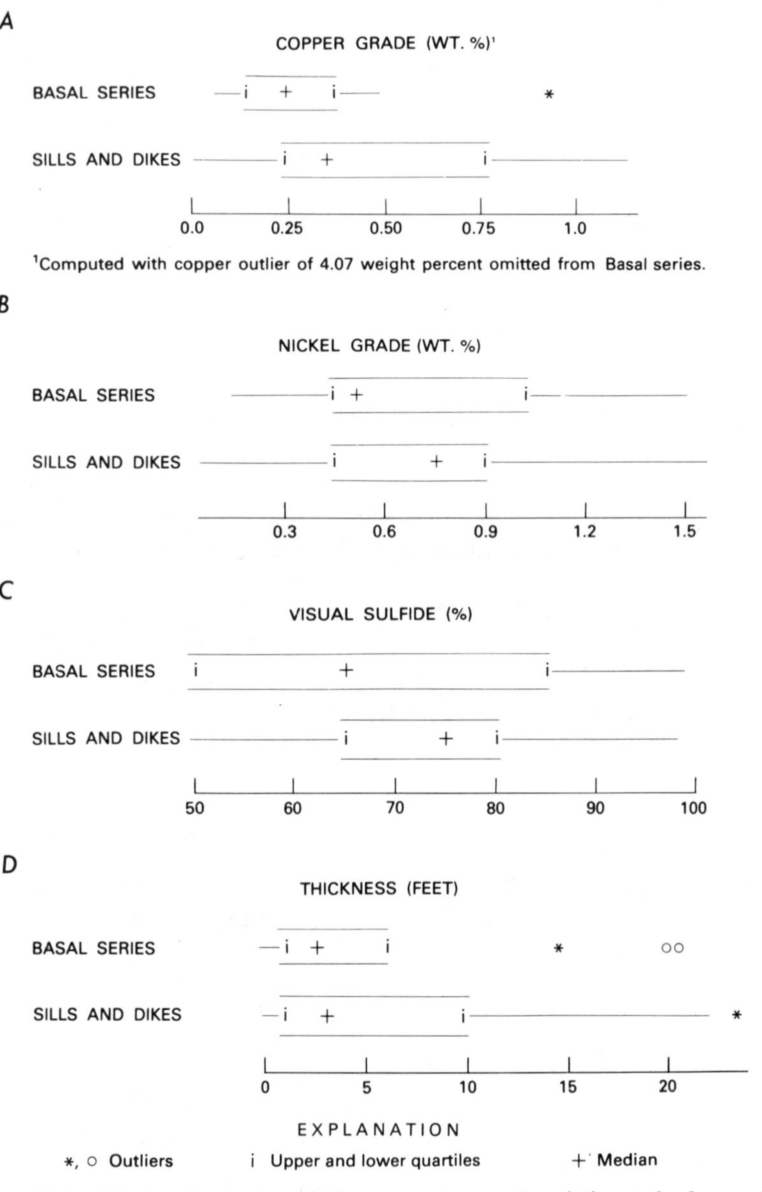

Figure 6.   Box plots showing (A) copper grade, (B) nickel grade, (C) percent visual sulfides, and (D) thickness of massive sulfides in Basal series and massive sulfides associated with sills and dikes of metasedimentary rocks.

Table 1. Summary statistics for massive sulfides in Basal series and massive sulfides associated with sills and dikes of metasedimentary rocks.

Unit		Copper* (wt,%)	Nickel (wt.%)	Visual Sulfide %	Thickness (feet)
Basal series	Mean	0.543	0.693	69.2	5.70
	Standard deviation	0.901	0.361	19.6	6.44
	Number of obs.	18.0	18.0	18.0	18.0
Sills and Dikes	Mean	0.514	0.756	73.5	7.28
	Standard deviation	0.315	0.367	13.7	7.58
	Number of obs.	25.0	25.0	25.0	25.0
P(w)	Significance levels	0.16	0.60	0.37	0.60

*With copper outlier removed

Unit		Copper* (wt,%)			
Basal series	Mean	0.335			
	Standard deviation	0.196			
	Number of obs.	17.0			
P(w)	Significance level	0.08			

These results were established by use of a Mann-Whitney-U test on these distributions (Table 1). The Mann-Whitney-U test is a nonparametric test based upon the sum of ranks assigned to the observations. It is less sensitive to outliers than the two-sample t-test. The relatively large copper mean and standard deviation in the massive sulfides of the Basal series as compared to massive sulfides associated with the sills and dikes are mainly the result of one outlier having a value of 4.07 weight percent copper. With this point removed, the copper mean and standard deviation are 0.335 and 0.196, respectively. The statistical test on all points, inclusive of the outlier, shows that the observed difference between the copper grades could occur by chance 16 percent of the time (P(w) = 0.16 for copper), or, with the outlier removed, 8 percent of the time. Similarly, the observed differences in the nickel grades could occur at random 60 percent of the time and those for visual sulfide 37 percent of the time; however, the equal variance assumption of this test seems to be violated. In addition to the Mann-Whitney-U test, the thicknesses of the massive sulfides in the Basal series were compared with those in the sills and dikes. From this comparison, we observe that the differences in the thickness distributions could have occurred 60 percent of the time by chance.

Therefore, on the basis of the Mann-Whitney-U test of the copper and nickel grades, we cannot reject (at the 5 percent significance level) the hypothesis that the massive sulfides in the Basal series and associated with the sills and dikes came from the same parent magmas.

(2) $H_0(2)$, the copper and nickel grades of the massive sulfides do not show a trend as a function of their distance above the base of the Basal series.

This would be expected because of the large amount of sulfides present, impulses of new magmas, possible convection processes resulting in magma mixing, and the possibility of postdepositional migration of sulfides within the Basal series, which would tend to homogenize grades.

In order to test this hypothesis, we plotted the copper and nickel grades of the massive sulfides occurring in the Basal series against their distance above the base of the Basal series contact. In Figure 7A, copper grade versus distance shows a range of copper grade from 0.13 to 0.95 percent within a true distance above the contact of 231 feet. As in Figure 6A and Table 1, the one large copper outlier (4.07 percent) is not shown. All but 5 of these 17 points are below 0.5 percent copper and 120 feet distance. A simple linear regression of copper grade upon distance with a sample size of 17 shows a Spearman's correlation of -0.042, which is not significant at the 5 percent level.

Figure 7B shows nickel grade versus distance above the base of the Basal series. Nickel ranges from 0.15 percent to 1.52 percent within a maximum true distance of 231 feet above the contact. The regression of nickel upon distance with a sample size of 18 shows a Spearman's correlation of 0.053, which is not significant at the 5 percent level. What can be seen is that all but two observations occur within 120 feet of the contact and all but one nickel grade is below 1.2 percent.

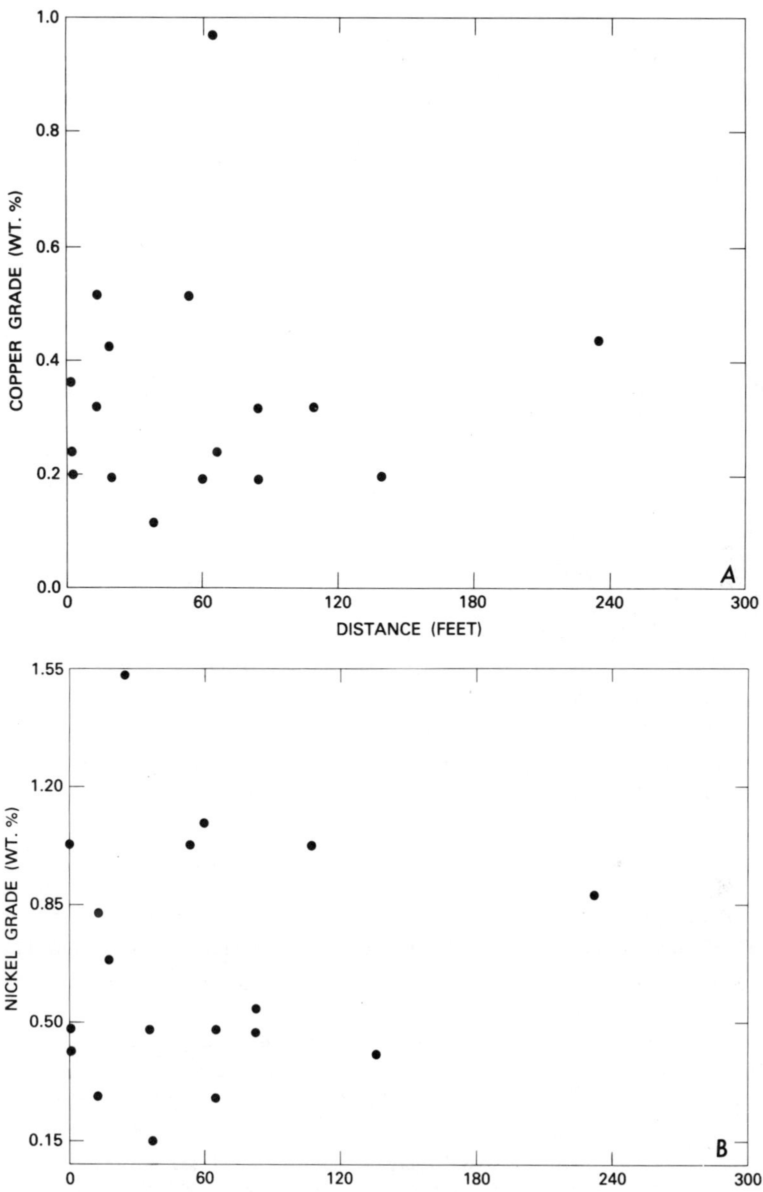

Figure 7. (A) Copper grade versus distance of massive sulfides above base of Basal series; (B) nickel grade versus distance of massive sulfides above base of Basal series.

This lack of association between copper grade versus distance and nickel grade versus distance above the base of the Basal series shows that grades of copper and nickel sulfide differ erratically, within 120 feet above the base of the Basal series.

These results suggest that copper and nickel concentrations in massive sulfides do not seem to be related to their distance above the base of the Basal series.

Therefore, we do not reject the hypothesis that copper and nickel grades do not show a trend as a function of distance above the base of the Basal series.

> (3) $H_0(3)$, the copper and nickel grades of the massive sulfides associated with the sills and dikes of the metasedimentary rock do not show a trend as a function of their distance below the base of the Basal series.

This result is expected because of a lack of vertical control on the distribution of the occurrence of massive sulfides. For this reason, these results are inconclusive as far as proving the sulfide liquid immiscibility model, but are important to the analysis of massive sulfides within the Mountain View area of the Stillwater Complex. Testing of this hypothesis was similar to that used for the previous hypotheses. A test of regression equations for data shown in Figures 8A and 8B did not indicate any linear trend at the 5 percent significant level in either copper (Spearman's correlation of 0.0349) or nickel (Spearman's correlation of -0.290) grades of the massive sulfide downward from the base of the Basal series. Grades for copper range from 0.10 to 1.15 percent, grades for nickel range from 0.06 to 1.56 percent, and the maximum true distance below the base of the Basal series was 158 feet for our data set. Thus, the pattern of variable copper and nickel grades within the massive sulfides, and the lack of association between copper or nickel grades and distance in the Basal series continues through massive sulfides associated with the sills and dikes.

Therefore, the hypothesis that the copper and nickel grades of the massive sulfides associated with the sills and dikes do not show a trend as a function of their distance below the Basal series is not rejected.

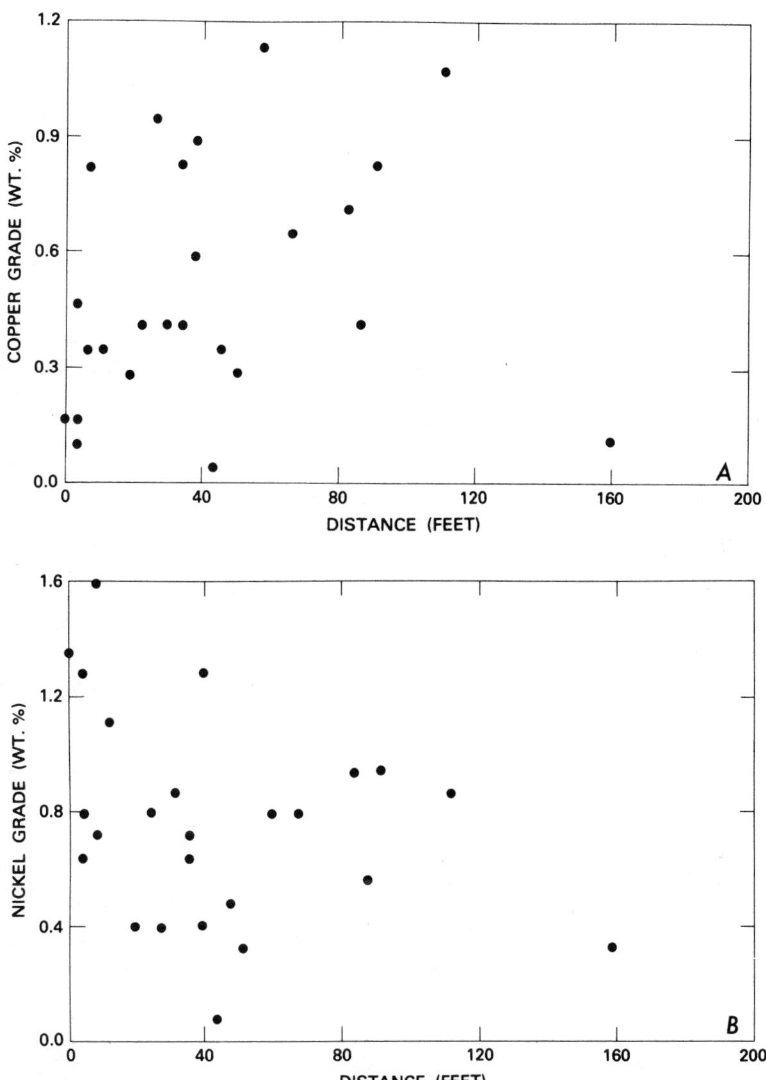

Figure 8. (A) Copper grade versus distance of massive sulfides below base of Basal series associated with sills and dikes in metasedimentary rocks; (B) nickel grade versus distance of massive sulfides below base of Basal series associated with sills and dikes in metasedimentary rocks.

> (4) $H_0(4)$, there is no significant correlation between the copper grade and thickness and nickel grade and thickness of the massive sulfide occurrences in the Basal series.

This is an expected result of the sulfide immiscibility model because of the large volume of sulfide present; convective processes differentially inhibiting and accelerating sulfide movement, which results in mixing; and remobilization and concentration of sulfides due to filter pressing of cumulate piles.

Scatterplots for the relationship between the copper and nickel grades and the thickness of the massive sulfides occurring in the Basal series are shown in Figures 9A and 9B, respectively. The intervals of sulfides in which visual sulfides are greater than 50 percent range in thickness from 0.4 feet to 21.1 feet and copper and nickel grades are the same as those in Figures 7A and 7B. Neither of the Spearman's rank correlation coefficients for the scatter plots shown in Figures 9A and 9B, $r_S = -0.257$ with a sample size of 17 and $r_S = -0.176$ with a sample size of 18, respectively, are significant at the 5 percent probability level.

Therefore, we conclude that the factors controlling the copper and nickel grades of the massive sulfide occurrences in the Basal series acted independently of the factors controlling their thickness. This result is an expected consequence of a liquid immiscibility model where mechanical mixing and remobilization is assumed to keep the sulfide magma homogeneous during both the emplacement and the accumulation processes.

> (5) $H_0(5)$, there is no significant correlation between the copper grade and thickness and nickel grade and thickness of the massive sulfide associated with the sills and dikes.

This result would be expected because the thicknesses and spatial distribution of the sills and dikes and associated massive sulfides are erratic with no known control, except to occur within 200 meters of the Basal series contact. As with $H_0(3)$, these characteristics are inconclusive as far as proving the sulfide immiscibility model, but are of interest in the analysis of massive sulfides of the Mountain View area.

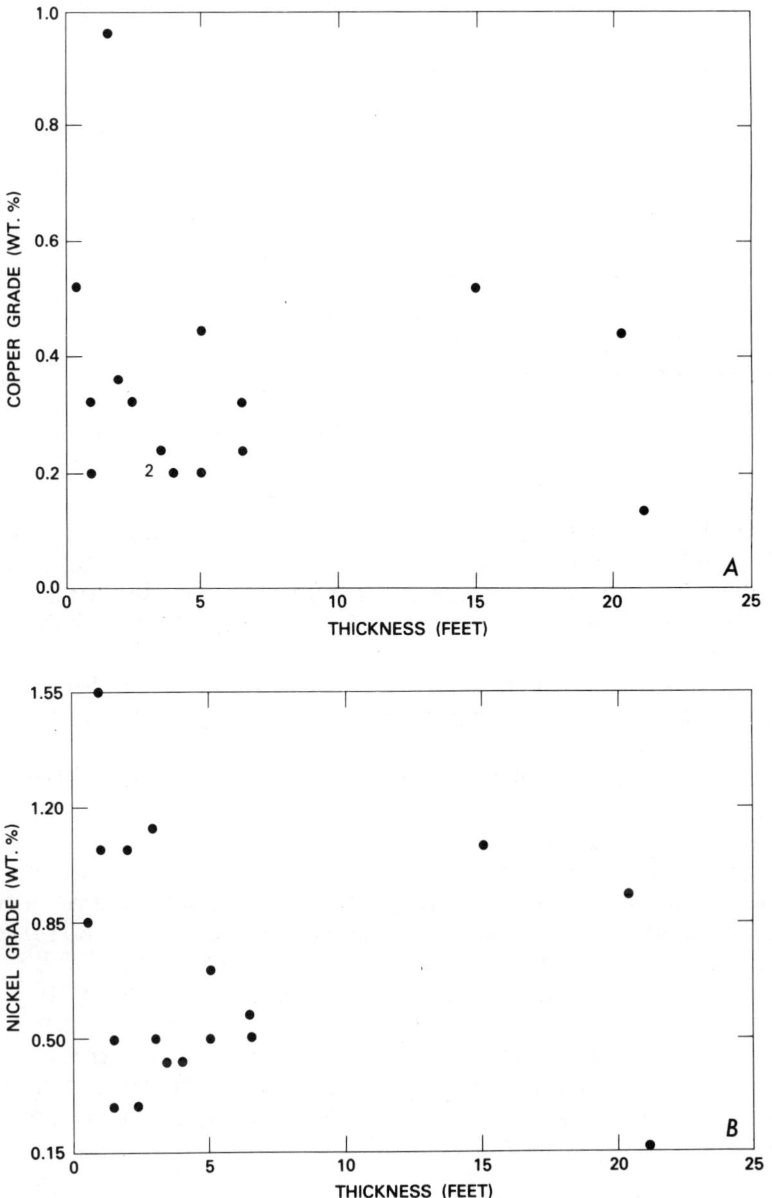

Figure 9. (A) Copper grade versus thickness of massive sulfides in Basal series; (B) nickel grade versus thickness of massive sulfides in Basal series.

Scatterplots analogous to those shown in Figures 9A and 9B for the Basal series are displayed in Figures 10A and 10B for massive sulfides associated with the sills and dikes. The thickness of the massive sulfide intervals in which visual sulfides are greater than 50 percent range from 0.3 feet to 24.3 feet, with copper and nickel grades the same as those shown in Figures 8A and 9B. The Spearman's rank correlation coefficient of -0.019 and -0.001 with a sample size of 25 for Figures 10A and 10B, respectively, are not significant at the 5 percent probability level.

Therefore, there is an absence of correlation between copper grade and thickness and nickel grade thickness of the massive sulfides associated with the sills and dikes of the metasedimentary rock.

(6) $H_0(6)$, the volume of the massive sulfides in the Basal series is greatest closest to the base of the Basal series.

This result would be expected because of the coalescing of sulfide droplets and gravitational migration towards the bottom of this unit. The volume of massive sulfides would be dependent on two variables: thickness and frequency of the massive sulfides.

A scatterplot showing the relationship between thickness of the massive sulfides in the Basal series and distance above the base of the Basal series is displayed in Figure 11A. A simple linear regression of thickness upon distance with a sample size of 18 results in a Spearman's correlation of 0.245, which is not significant at the 5 percent probability level. The scatterplot and regression results suggest that the process that accumulates the sulfide liquid does not leave any pattern in thickness as a function of distance above the base of the Basal series. Although the massive sulfides in the Basal series do not become thicker closer to the base of the Basal series, evidence suggests that they do become more abundant (Fig. 11B). Using 25-foot intervals away from the base of the contact and accumulating thicknesses by 25-foot intervals, this sum of total sulfides intercepted in the Basal series shows an increase as the base of the Basal series is approached.

Possibly the massive sulfides do not increase in thickness downward in the Basal series because the silicate-crystal-accumulation rate may have been fast enough to prohibit the sulfides from coalescing into thick bodies. The

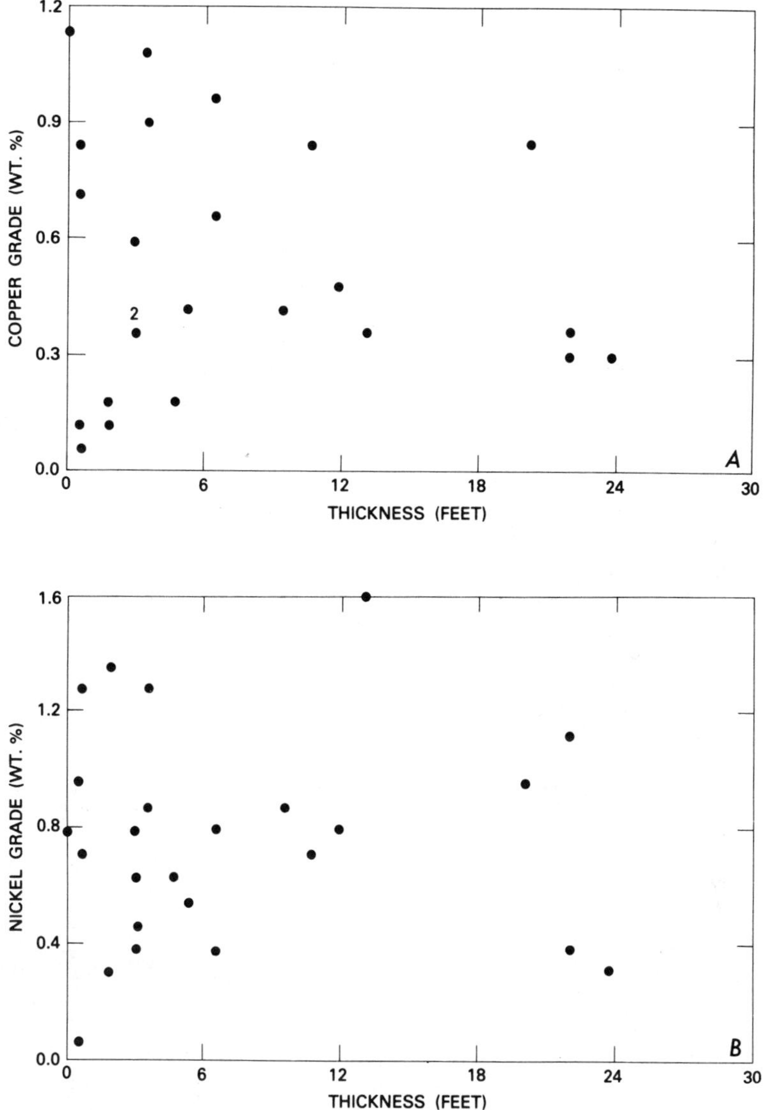

Figure 10. (A) Copper grade versus thickness of massive sulfides below base of Basal series associated with sills and dikes in metasedimentary rocks; (B) nickel grade versus thickness of massive sulfides below base of Basal series associated with sills and dikes in metasedimentary rocks.

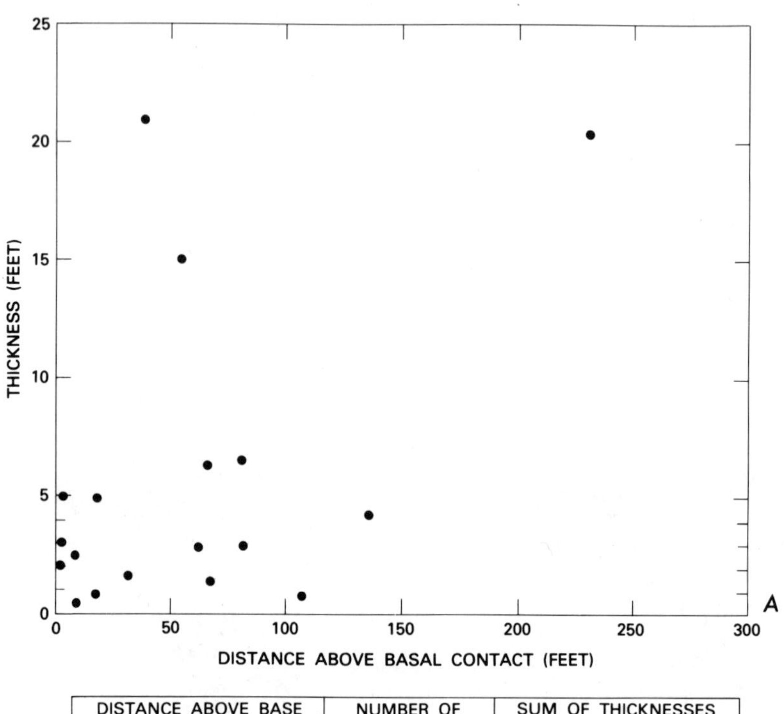

DISTANCE ABOVE BASE OF BASAL SERIES (FEET)	NUMBER OF OBSERVATIONS	SUM OF THICKNESSES WITHIN INTERVAL (FEET)
0–25	7*******	19.12
25–50	2**	22.77
50–75	4****	25.77
75–100	2**	9.41
100–125	1*	0.85
125–150	1*	4.23
150–175	0	—
175–200	0	—
200–225	0	—
225–250	1*	20.48

Figure 11. (A) Thickness versus distance of massive sulfides above base of Basal series; (B) Frequency of occurrence by 25-foot intervals of massive sulfides above base of Basal series.

suggestion of an increase in frequency and, therefore, massive sulfide volume is viewed as support for this model.

(7) $H_O(7)$, the volume of massive sulfides associated with the sills and dikes within the metasedimentary rocks is greatest closest to the base of the Basal series.

This result would be expected in the liquid immiscibility model because areas proximal to the zone of accumulation should contain more sulfide than areas distal to the zone of accumulation. An association has been determined between the occurrence of massive sulfides and adjacent sills and dikes within the metasedimentary rocks (Zientek, 1983). The sills and dikes associated with the Basal series are assumed to increase in frequency as distance from the Basal series contact decreases.

The data used to test this hypothesis are displayed in Figure 12A and 12B. A simple linear regression of thickness upon distance from the base of the Basal series and into the associated sills and dikes of the metasedimentary rock as shown in Figure 12A (Spearman's rank correlation coefficient of -0.090 with a sample size of 25), is not significant at the 5 percent probability level. This result suggests that the massive sulfides associated with the sills and dikes do not become thinner below the base of the Basal series. However, Figure 12B shows rather clearly that these massive sulfides do become less frequent with increasing distance below the base of the Basal series. Therefore, the volume of sulfide becomes less as distance from the contact increases.

This decrease in the volume of sulfide away from the base of the Basal series is expected in the sulfide liquid immiscibility model. However, thickness, frequency, and distribution of the massive sulfides in the metasedimentary rock are controlled by characteristics of the host metasedimentary lithology, independent of sulfide liquid immiscibility.

(8) $H_O(8)$, the copper grades are correlated with the nickel grades of the massive sulfides in the Basal series.

The relationship between the copper and nickel grades in the massive sulfides intersected in the Basal series is shown in Figure 13. This figure reiterates what we have seen already in Figure 6 and Table 1: proportionally, more nickel than copper is present

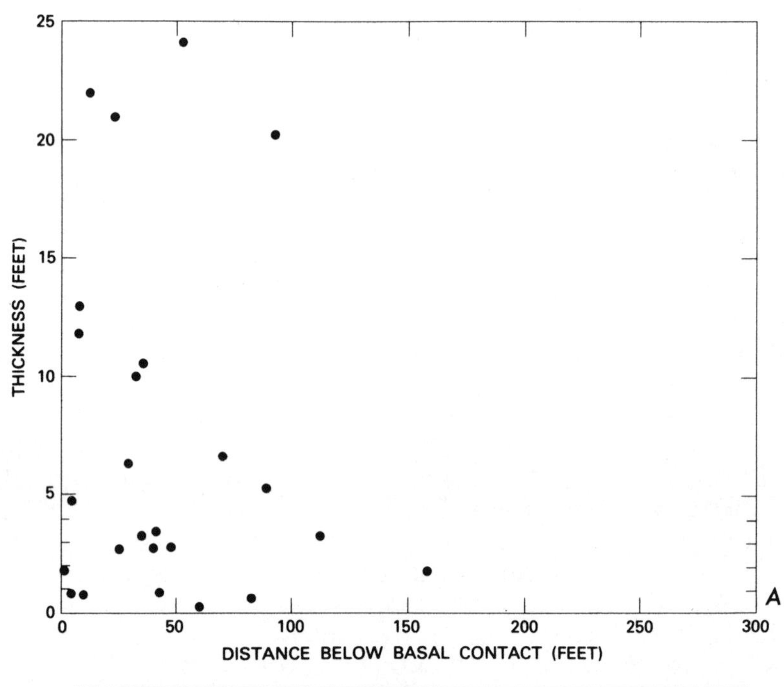

DISTANCE BELOW BASE OF BASAL SERIES (FEET)	NUMBER OF OBSERVATIONS	SUM OF THICKNESSES WITHIN INTERVAL (FEET)
0–25	8********	77.07
25–50	9*********	42.45
50–75	3***	31.17
75–100	3***	26.25
100–125	1*	3.38
125–150	0	—
150–175	1*	1.69

Figure 12. (A) Thickness versus distance of massive sulfides below base of Basal series associated with sills and dikes in metasedimentary rocks; (B) Frequency of occurrence by 25-foot intervals of massive sulfides below base of Basal series associated with sills and dikes in metasedimentary rocks.

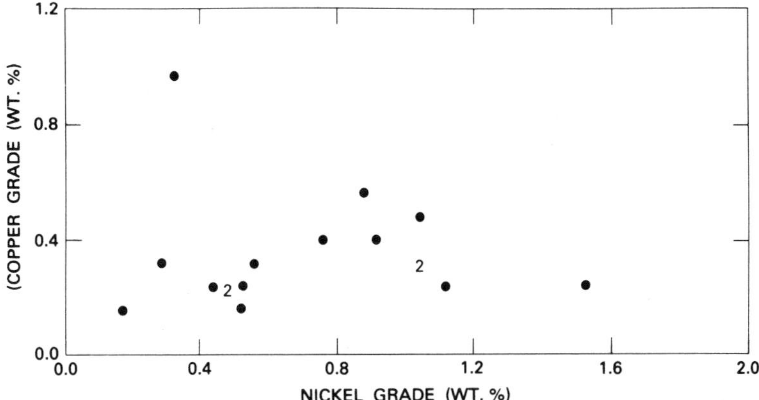

Figure 13. Copper grade versus nickel grade of massive sulfides in Basal series.

in the massive sulfides of the Basal series and the nickel has a wider range of values (excluding one copper outlier of 4.07%). The Spearman's rank correlation coefficient for this relationship is r_S = 0.126 with a sample size of 17, which is not significant statistically at the 5 percent level. The null hypothesis, therefore, is rejected and it is concluded that the copper and nickel grades do not covary positively within the massive sulfides in the Basal series.

(9) $H_0(9)$, the copper and nickel grades are correlated in the massive sulfides associated with the sills and dikes in the metasedimentary rock.

The scatterplot for the copper and nickel grades of the 25 massive sulfide intersections associated with the sills and dikes is shown in Figure 14. The Spearman's rank correlation coefficient for this relationship is r_S = 0.119, which is not significant at the 5 percent level. The null hypothesis therefore is rejected and it is concluded that the copper and nickel grades within the massive sulfides in this unit are not correlated.

It must be pointed out, however, that this lack of correlation between copper and nickel grades in the massive sulfides of the Basal series and the massive sulfides associated with the sills and dikes of the metasedimentary rocks is not true for less concentrated sulfides within these units. For example, in another

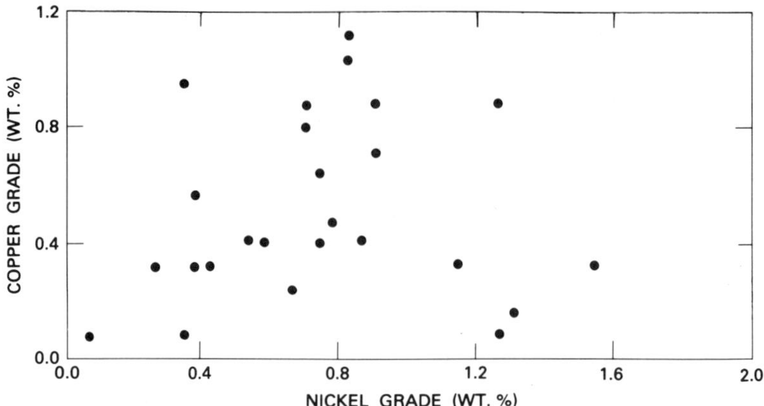

Figure 14. Copper grade versus nickel grade of massive sulfides associated with sills and dikes in metasedimentary rock.

study an analysis of all intervals greater than or equal to 5 percent visual sulfides within the Basal series for copper versus nickel yielded a Pearson's correlation coefficient of 0.447 and a Spearman's correlation coefficient of 0.214 with a sample size of 82. An analysis of this same range of sulfides associated with sills and dikes shows a Pearson's correlation of 0.330 and Spearman's correlation of 0.432 with a sample size of 134. These correlation coefficients show a statistical correlation between copper grades and nickel grades in intervals of greater than 5 percent visual sulfides for both the Basal series and the sulfides associated with the sills and dikes.

This suggests that the sulfides present in concentrations greater than 50 percent apparently differ from sulfides present in concentration of less volume. This is indicative that massive sulfides may be one end-member of a process or decoupled from silicate magma at some point in the development of the cumulate member.

Most of the statistical tests lead to an acceptance of the null hypothesis. Because the samples generally were small, rejecting the null hypotheses is difficult, and the tests have minimal power. Accepting the null means evidence is insufficient to reject the hypothesis. Thus, the probability of a type II error, accepting a hypothesis which is incorrect, may be large. Another concern is that of multiple comparisons. When a number of comparisons are made, each at a given significance level α, the probability of making at least one type I error, that is rejecting a hypothesis which is correct, is greater than α. In spite of these concerns which should lead to a cautious interpretation of the result of this study, a methodology has been demonstrated which allows a geological model to be tested using statistical procedures.

CONCLUSIONS

The results of tests of nine hypotheses have led to the conclusion that there is no evidence to reject the liquid immiscibility model as providing the required mechanisms to transport and emplace the copper- and nickel-rich massive sulfides that are near the base of the Stillwater Complex. The nine hypotheses tested were based upon expectations derived from assumptions about the source and genesis of the sulfides, mode of introduction of the sulfides, the partitioning of the copper and nickel into the sulfide phase at the time of magma crystallization, the mechanical mixing of the silicate-sulfide magma during transport, and emplacement of the sulfide phase. The hypotheses tested included expectations about the distribution of the copper and nickel grades of the massive sulfides which occur in the Basal series, the existence or nonexistence of trends in the thickness, frequency, grade, and volumes of the massive sulfides with respect to stratigraphic position, and the nature of the correlation structure between the copper and nickel grades of these massive sulfides. Specific results include:

(1) The massive sulfides in both the Basal series and associated with sills and dikes of the metasedimentary rocks are similar statistically with respect to both the copper and nickel grades.

(2) The copper and nickel grades of massive sulfides are not related to their thicknesses in either Basal series or the associated sills and dikes of the metasedimentary rocks.

(3) No trend in thickness of the massive sulfides was determined as a function of their stratigraphic position in either the Basal series or the associated sills and dikes of the metasedimentary rocks.

(4) Evidence was determined for a declining frequency of occurrence of massive sulfides as a function of distance away from the base of the Basal series in both the Basal series and metasedimentary rocks.

This result is assumed to be a consequence of the coalescing and gravitational migration of sulfides in the Basal series and the higher frequency of occurrence of sills and dikes near the top of the metasedimentary rocks.

(5) The total volume of sulfide, which is a consequence of massive sulfide thickness and frequency, increases with depth within the Basal series and decreases with distance from the base of the Basal series within the metasedimentary rock interval intruded by sills and dikes.

(6) Although copper grades and nickel grades are known to exhibit a systematic change in sulfides as a function of distance in some mafic intrusions, no association in grades of massive sulfide versus distance could be recognized.

REFERENCES

Barton, P.B., Jr., 1970, Sulfide petrology: Mineral. Soc. America Spec. Paper 3, p. 187-198.

Bawiec, W.J., 1985, Computer applications to structural interpretation and metal distribution within the Basal series of the Stillwater Complex, Montana: EOS, v.66, no. 18, p. 398-399.

Bawiec, W.J., and Drew, L. J., 1984, Basal zone of the Stillwater Complex, Montana: Internal stratigraphy and morphology from copper and nickel assay values (abst.): Geol. Soc. America, Reno, NV, Nov., 1984, p. 508.

Cowan, J.C., 1968, The geology of the Strathcona ore deposit: Canadian Mining and Metall. Bull., v. 1, no. 669, p. 38-54.

Craig, J.R., and Kullerud, G., 1969, Phase relations in the Cu-Fe-Ni-S system and their application to magmatic

ore deposits, in Wilson, H.D.B., ed., Magmatic ore
deposits: Econ. Geology Mon. 4, p. 344-358.

Drew, L.J., Bawiec, W.J., and Page, N.J. 1983, The copper-nickel
assay log: A tool for stratigraphic interpretation
within the Basal zone of the Stillwater Complex (abst.):
EOS, v. 64, no. 45, p. 884.

Drew, L.J., Bawiec, W.J., Page, and N.J., Schuenemeyer, J.H.,
1985, The copper-nickel concentration log: A tool for
stratigraphic interpretation within the Ultramafic and
Basal zones of the Stillwater Complex, Montana: Jour.
Geochem. Expl., v. 23, no. 2 p. 117-137.

Genkin, A.D., Eustigheeva, T.L., and Kovalenker, V.A., 1980,
Platinum group minerals of copper-nickel ores and
some aspects of their origin: Proc. 11th Intern.
Mineralog. Assoc. Congress (Novosibirsk), p. 165-171.

Haughton, D,R., Roeder, P.L., and Skinner, B.J., 1974, Solubility of
sulfur in mafic magmas: Econ. Geology, v. 69, no.4,
p. 451-467.

Hawley, J.E., 1962, The Sudbury ores - their mineralogy and
origin: Can. Mineralogist, v. 7, pt. 1, p. 1-207.

Hawley, J.E., 1965, Upside-down zoning at Frood, Sudbury,
Ontario: Econ. Geology, v. 60, no. 3, p. 529-575.

Hess, H.H., 1960, Stillwater igneous complex, Montana - a
quantitative mineralogical study: Geol. Soc. America
Mem. 80, 230 p.

Jackson, E.E., 1961, Primary textures and mineral association in
the ultramafic zone of the Stillwater Complex,
Montana: U.S. Geol. Survey Prof. Paper 358, p. 1-106.

Jones, W.R., Peoples, J.W., and Howland, A.L., 1960, Igneous and
tectonic structures of the Stillwater Complex,
Montana: U.S. Geol. Survey Bull. 1071-H, p. 281-340.

McCallum, I.S., Raedeke, L.D., and Mathez, E.A., 1980,
Investigations of the Stillwater Complex: Part I.

Stratigraphy and structure of the Banded zone: Am. Jour. Sci., v. 280-A, pt. 1, p. 59-87.

Naldrett, A.J., 1981, Nickel sulfide deposits: Classification, composition, and genesis: Econ. Geology 75th Ann. Vol., p. 628-685.

Page, N.J., and Nokelberg, W.J., 1974, Geologic map of the Stillwater Complex, Montana: U.S. Geol. Survey Misc. Geol. Inv. Map I-797, 5 sheets, scale 1:12,000.

Page, N.J., 1977, Stillwater Complex, Montana - rock succession, metamorphism, structure of the complex and adjacent rocks: U.S. Geol. Survey Prof. Paper 999, 79 p.

Page, N. J. 1979, Stillwater Complex, Montana - structure, mineralogy, and petrology of the Basal zone with emphasis on the occurrence of sulfides; U.S. Geol. Survey Prof. Paper 1038, p. 1-69.

Page, N.J. Zientek, M.L., Czamanske, G.K., and Foose, M.P., 1985, Sulfide mineralization in the Stillwater Complex and underlying rocks, in Czamanske, G.K., and Zientek, M.L., eds., Stillwater Complex, Montana: Geology and guide, Montana Bureau of Mines and Geology, Spec. Publ. 92, p. 93-96.

Segerstrom, K., and Carlson, R.R., 1982, Geologic map of the Banded zone of the Stillwater Complex and adjacent rocks, Stillwater, Sweetgrass, and Park Counties, Montana: U.S. Geol. Survey Misc. Invest. Series Map-I-1383, 2 sheets, 1:24,000.

Skinner, B.J., and Peck, D.L., 1969, An immiscible sulfide melt from Hawaii, in Wilson, H.D.B., ed., Magmatic ore deposits: Econ. Geology Mon. 4, p. 310-322.

Todd, S.G., Keith, D.W., LeRoy, L.W., Schissel, D.J., Mann, E.L., and Irvine, T.N., 1982, The J-M platinum-palladium reef of the Stillwater Complex, Montana: I. Stratigraphy and petrology: Econ. Geology, v. 77, no. 6, p. 1454-1480.

Zientek, M.L., 1983, Petrogenesis of the Basal zone of the
        Stillwater Complex, Montana: unpubl. doctoral
        dissertation, Stanford Univ., 229 p.

Zientek, M.L., 1985, Czamanske, G.K., and Irvine, T.N., 1985,
        Stratigraphy and nomenclature for the Stillwater
        Complex, in Czamanske, G.K. and Zientek, M.L., eds.,
        Stillwater Complex, Montana: Geology and guide,
        Montana Bureau of Mines and Geology, Spec. Publ. 92,
        p. 21-32.

COMPUTER PROCESSING OF DIPMETER LOG DATA: ENHANCEMENT OF A SUBSURFACE EXPLORATION TOOL

Katherine J. Beinkafner

Casper, Wyoming

ABSTRACT

The dipmeter is unique in that it is the only logging tool that provides information describing conditions outside the borehole. Reliable predictions of the structural configuration in surrounding rock units can be made by extrapolation of proper three-dimensional interpretations from computer processed results. The three-dimensional conceptualization required to decipher structural and stratigraphic information from dipmeter logs can be aided superlatively by the use of computer-generated displays, generally not provided by the well-logging service companies.

The significance of further manipulation of tadpole (arrow plot) data is emphasized. The variable components, graphic format, and geologic significance is demonstrated for the following plots: continuous azimuth vector plot, borehole deviation block diagram, fracture identification plot in azimuth space, SCAT (Statistical Curvature Analysis Techniques expounded by C. A. Bengtson of the USA) plots, and cross with which the borehole path dip components, formation boundaries, and structural axes are projected onto a vertical cross section of any orientation either intersecting the borehole or some distance away, perhaps along a seismic line. Production of these simple graphic illustrations is a task computers have been programmed to do quickly and unerringly. Once available these transformations save time, they provide a simple tool for interpretation, and they reveal the implicity of complex geologic terrains.

INTRODUCTION

In the petroleum industry the dipmeter well-logging tool is used to determine the magnitude and direction of dip of strata penetrated by the borehole. The dipmeter tool consists of four arms that make contact with the wall of the borehole as the tool is raised to the surface. Microresistivity of the rock wall is measured at a constant sampling rate usually every 0.2 or 0.1 inch (5.08 or 2.54 mm). The earliest tools had three arms for minimal definition of a plane by correlation of the microresistivity curves. The orientation of the tool relative to horizontal and magnetic north also is recorded, although at a somewhat slower rate of every 3.2 inches (8.128 cm). Two caliper curves measuring the separations of pairs of arms 1-3 and 2-4 are recorded at the same rate.

The geometry of separate curves measured around the borehole allows the determination of the orientation of planar features that the borehole penetrates (Fig. 1). The most obvious features detected by correlation are sedimentary strata, and it may be possible to detect fractures, joints, faults, diagnetic layering, and fluids in pore space.

The service companies generally provide two basic graphic plots of dipmeter data. The field monitor log displays the microresistivity curves and the tool orientation curves (azimuth, relative bearing, and borehole deviation) in log format (Fig. 2). The tadpole or arrow plot was devised to display the three-dimensional variations of dip with depth in two-dimensional pace. At each log depth the best correlation of the microresistivity curves defines a plane. A tadpole at that depth (Fig. 3) represents the magnitude and direction of planar dip. The position of the tadpole body from let to right indicates the magnitude of dip (0 to 90 degrees), whereas the tail points along the downdip azimuth. Azimuth is equivalent to compass directions with north to the top of the log, south toward the bottom, east to the right and west to the left.

The four arm tools make one measurement per arm with an extra electrode placed a short distance vertically above the other on one

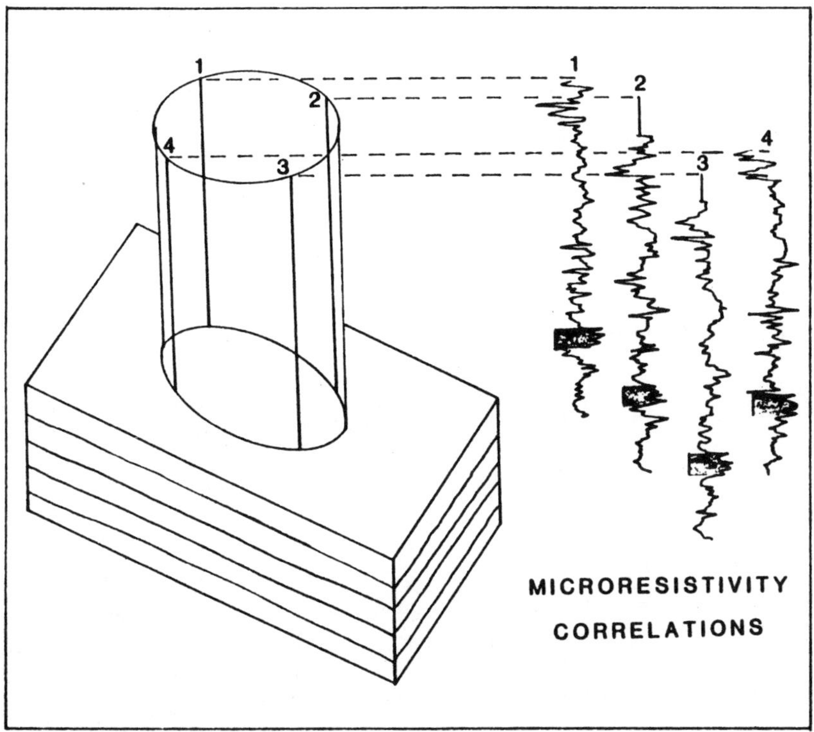

Figure 1.   Schematic diagram of well penetrating sedimentary
beds. Four microresistivity curves measured along
borehole are correlated in shaded zone to determine
dip of strata.

arm to use for speed corrections. By correlation of the two curves
from the same arm, variations in uphole velocity can be detected
and the digital microresistivity curves can be stretched or
condensed to remove the effects of tool sticking, jutter, and bob.
Some new tools have two side-by-side electrodes on each arm and
another has six arms. The trend in tool development is to collect
more data in time and space to achieve the objective of finer
resolution of microresistivity features. Speculation is that a
change in the rock property that is measured may be the next
innovation. Future tools may employ high resolution sonic,
density, or video recording in the multiple arm geometry.

With the three arm tool used in the 1960's, the correlations
between pairs of microresistivity curves were done optically with

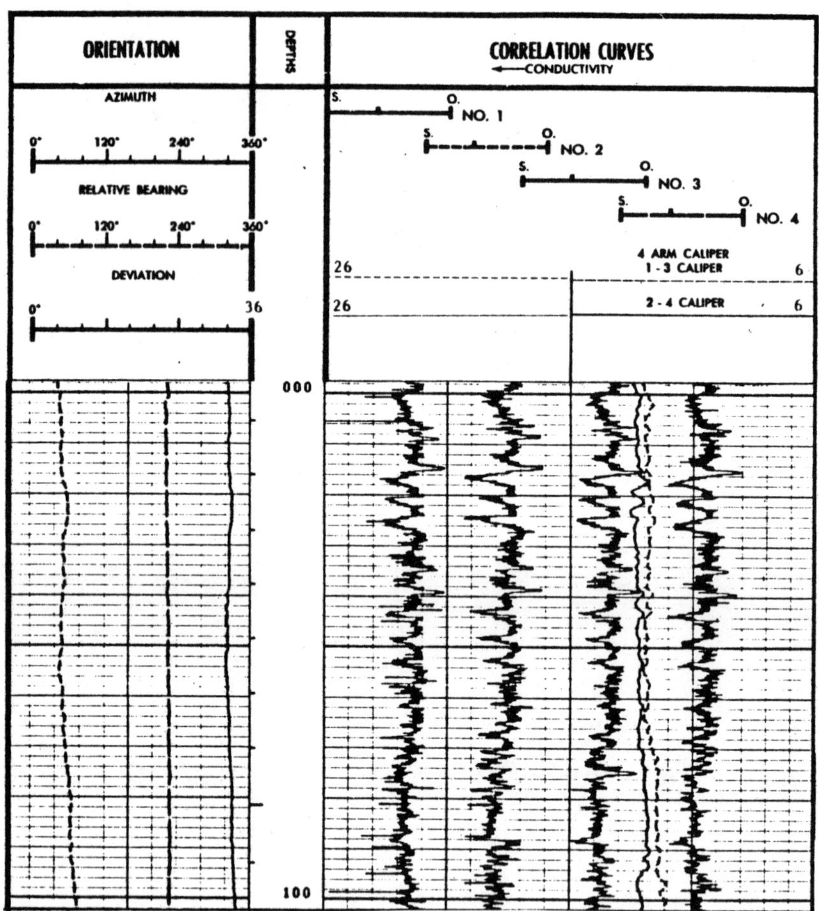

Figure 2. Field monitor log with orientation, microresistivity, and caliper curves.

a comparator. With increases in sampling rate and the advent of computers, the correlations are done by computers. The frequency of correlations with depth is selected by the geologist for computer processing by the service company or by the oil company. Some oil companies and consulting firms have programs for reading the dipmeter field tapes, computing correlations, and displaying results. The three basic parameters controlling dip computation are correlation window, step, and

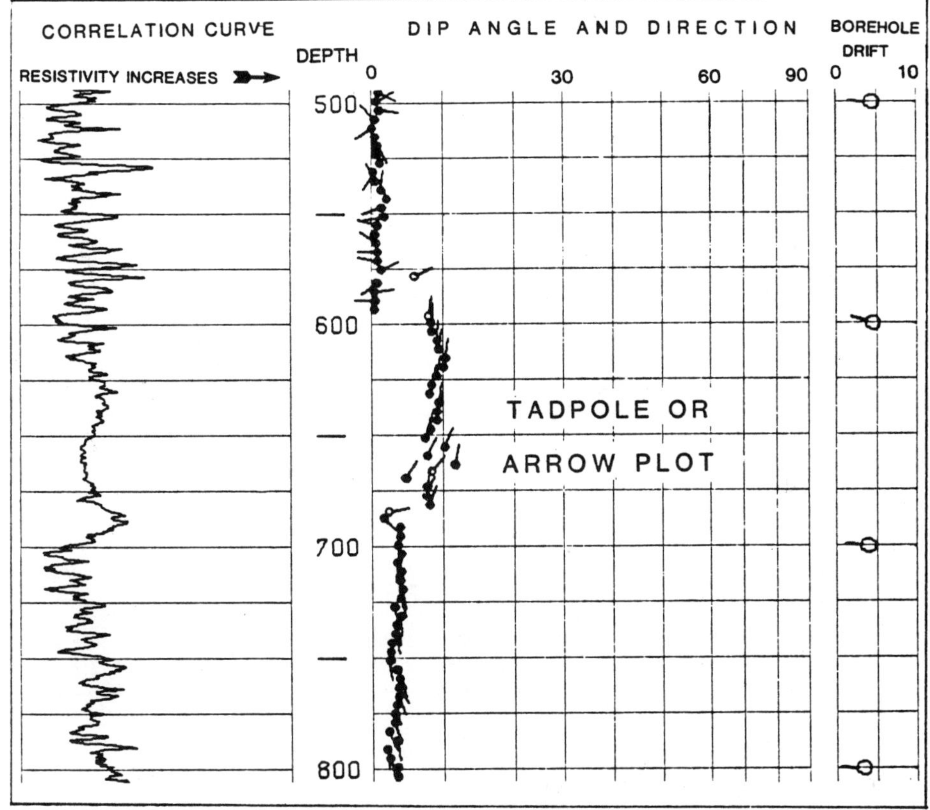

Figure 3. Tadpole or arrow plot.

search angle (Fig. 4). The correlation window is the vertical length of microresistivity curve that is selected and compared for each determination. The step is the distance that the window is moved along the microresistivity curves for the next correlation or the vertical distance between successive tadpoles. The search angle is proportional to the distance up or down the well that the curves are searched for the best correlation.

The quality of correlations is judged by computation of planarity, closure, and likeness. Closure is the sum of the four individual vertical shifts in the correlation of pairs of adjacent neighbor curves around the borehole. Closure error is the sum of the signed (positive and negative) pair shifts divided by the sum of their absolute values. Zero closure indicates that all upward shifts

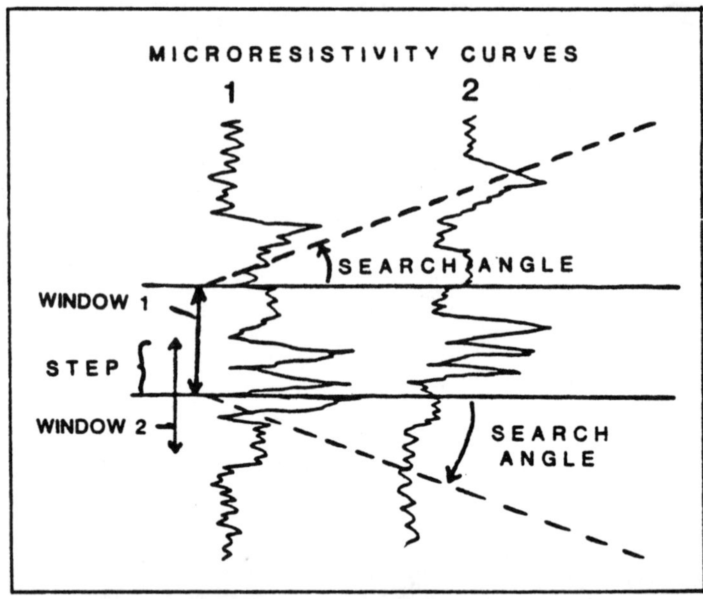

Figure 4. Three parameters (window, step, and search angle) used in computer correlations between pairs of microresistivity curves.

are compensated by downward shifts. Planarity is a measure of how well the selected correlation defines a plane because it is overdetermined. This quality is measured by the vertical distance between two perpendicular lines connecting correlation points on opposing curves 1 with 3 and 2 with 4. Likeness is the lowest correlation coefficient of the six possible pairs of correlations. These quality measurements are useful in the process of eliminating noise from the data.

In spite of the technical advances in tool design and data acquisition by the service companies, the interpretive techniques for defining geologic structures has been developed slowly by geologists. The reasons for this interpretational lag are complex. A description of the empirical evolution of the analytical methods may explain some of the problems encountered in the development stages.

EVOLUTION OF INTERPRETATIVE TECHNIQUES

The capability of the dipmeter to predict the geometry of rock strata away from the borehole makes it an exploration tool rivalled only by seismic surveys at much greater expense. The fact remains that the dipmeter log is regarded as an orphan and is left to fall into the abyss between the disciplines of geology and well-log analysis. Some introductory questions that may lead to an understanding of how this unfortunate fate has befallen the dipmeter include:

(1.) Why are the results considered suspect by many geologists?

(2.) Why are the logs not interpreted fully to live up to their potential?

Upon consideration of the quality of results, it is important to know what the tool actually measures. The tool is reported to measure changes in electrical microresistivity of the rocks encountered in the borehole. Studies have shown that dips of bedding plane measurements from core and dipmeter logs may not match over a specific interval (Rider, 1978). These inconsistencies may arise for many reasons. The core may not be matched precisely in depth and orientation with the dipmeter log. The strata may not have microresistivity contrasts. The dipmeter may be detecting changes in permeability, fluid content, grain size, chemical content of the rock matrix, or a combination of these factors. The geometric distribution of these characteristics may or may not be parallel to the bedding planes of sedimentary rocks.

Another consideration is scatter of the dipmeter data. Examination of a tadpole plot reveals scatter of dips in both magnitude and direction. Geologic processes influence grain-size distribution and sorting in sedimentary rocks. Generally marine shales have fairly consistent dips and high-quality correlations, whereas conglomerates exhibit a greater scatter in dip azimuth and magnitude from lower quality correlations. Sandstones and siltstones are intermediate in this spectrum of data scatter. Because sedimentation is a random process, particle placement, bed thickness, and bedding plane orientation also may have a random distribution about a man. Individual beds may exhibit a trend in azimuth and magnitude of dip, reflecting the depositional gradient, but there usually is plenty of scatter of individual data points on tadpole plots. Hence, significant results are derived from trends of many tadpoles, rather than from individual tadpoles.

The volume of dipmeter data is another inherent characteristic that leads to complications in data processing, storage, and interpretation. With microresistivity measurements taken on five or more electrodes every 0.2 inch (5.08 mm) and tool orientation and calipers every 3.2 inches (8.128 cm), there are a total of 320 values per foot (1053 per meter) compared to two measurements per foot (7 per meter) on a conventional log, such as gamma ray, neutron, density, or sonic. This quantity of data may require offline magnetic tape storage and special facilities to process such large arrays. The correlation computations are time consuming and expensive. Data storage, software development, and reprocessing are expensive tasks, but the resultant reinterpretation may be well worth the cost.

Service contractors presented the first techniques of interpretation using pattern recognition by visual inspection of tadpole plots. Identification of folds, faults, and unconformities is successful with this method providing the structure is simple and has low plunge and borehole deviation is minimal. These exemplary methods do not account for the reality that dipmeter interpretation may involve unravelling the combined vectoral effects of

- directional drilling of a deviated borehole,
- regional dip,
- presence of structural elements, maybe overlapping,
- plunge of structures,
- multiple phases of deformation,
- multiple axes of deformation,
- geometric descriptions of folds, faults, or
  unconformities, and
- compressional versus extensional tectonic settings.

Bengtson (1980a, 1981, 1982) has shown that by employing some knowledge of the basics of structural geology, a few simple plots can be used to interpret complex geologic structures. Bengtson uses the concept that all geologic structures have unique transverse (T) and longitudinal (L) directions.

The transverse (T) direction is the direction of greatest variation of dips. A transverse cross section displays the displacement of

faults and curvature of folds. On maps the T-direction is parallel
to the maximum dip or perpendicular to the contour lines. The
longitudinal (L) direction is the direction of least variation in dips
and is perpendicular to the transverse direction. T- and
L-directions are defined for stratigraphic, structural, and
unconformal situations. The stratigraphic T-direction is in the
direction of regional dip or specifically may refer to the direction
of maximum dip through an interval of cross bedding, channel fill,
or prograding delta. The structural L-direction may be defied by
fold axes, fault strike, or other structural elements. The
transverse cross section of an angular unconformity shows the
maximum angular separation of the overlying and underlying beds.
A combination of T-directions may apply to he strata penetrated
by a particular well, depending upon the configuration
encountered.

The decomposition of dip vectors into directional components
can be summed over all depths for each direction on the compass
face. The resultant distribution of dip components can be used to
define the transverse and longitudinal directions. The greatest
variability of dip components occurs in the transverse direction
and the least in the longitudinal direction.

Once the T- and L-directions are known, each dip can be
decomposed into its vectoral components in these two
perpendicular directions. These dip components can be plotted
with depth on T-plots and L-plots. Given information on whether
the tectonic setting is compressional or extensional, the patterns
on SCAT plots are diagnostic of specific geologic structures. SCAT
is the acronym for Statistical Curvature Analysis Techniques
originated by Bengtson (1980a, 1981, 1982). The important
point here is that a geologic basis is used to display the data and
select the appropriate structural interpretation from a myriad of
combinations. A few geometric parameters can be read directly
from the SCAT plots to define the physical orientation of the
structures present (for example, plunge, T-direction, regional
dip). An additional construction may be necessary to determine a
vectoral combination, for example, determining an unconformity
on a tangent plot (Bengtson, 1980b, 1983).

Bengtson's innovations have improved upon the pattern
recognition concept of interpretation by providing a geologic
framework within which to work. Bengtson's methods can be
applied to the entire spectrum of deformational and sedimentary
structures, whereas previous authors have limited their

perspective to one type of feature or geographic area, such as reefs or the Gulf Coast.

Working independently in the United Kingdom, Shields (Z and S Consultants Limited, London) devised a computer program by the name of INCLINE that reads magnetic tapes of dipmeter data from service companies, processes the data using original algorithms, and displays the results in graphic formats.

Field measurements from magnetic tape provided by the service contractor are read into computer files with the INCLINE program. The "fast" file contains the microresistivity measurements which are recorded "fast" in the logging run. The "slow" file stores the caliper and orientation data which is recorded less frequently at the "slow" sampling rate.

Correlations of the microresistivity curves are performed and dips are computed using the caliper and orientation data. The results of depth, dip, and azimuth are recorded in a "correlation" file with a correlation associated with each depth. The correlation coefficients are similar to the Pearson product moment correlation coefficient in that they range from -1.0 to +1.0. This coefficient then can be used to select high-quality data for further plotting and analysis. INCLINE provides the basic plots of microresistivity, orientation, caliper or tadpole data at any specified scale.

As a consequence of programming and interpreting dipmeter data, Shields has created some graphic displays that provide geologic and geometric insight necessary for interpretation. The continuous azimuth vector plot, the borehole deviation survey block diagram, and the oriented microresistivity anomaly log in azimuth space are examples of his work that will be shown later.

Given the preceding historical and technical background, it is appropriate to investigate the use of computer-generated graphic tools for dipmeter analysis.

COMPUTER-GENERATED GRAPHIC AIDS FOR INTERPRETATION

Several computer-generated graphic displays are helpful in decomposing the vectors and visualizing the three-dimensional configuration of strata around the borehole. A scheme of

Table 1. Steps in geologic interpretation of dipmeter logs.

1. Determine how many vertical zones exist on the log with
    different T- and L-directions by detecting unconformities
    or detachments on continuous azimuth vector plot or SCAT
    plots.

2. Estimate the T-direction from tadpole plots and continuous
    azimuth vector plot.

3. Create SCAT plots: DVA plot, A-plot, D-plot, T-plot, L-plot.

4. By examining the DVA plot:
    a. Identify patterns of general structural setting.
    b. Estimate the regional dip.
    c. Determine T-direction and compare with step 2.
    d. Estimate the plunge.

5. On the SCAT depth plots:
    a. Draw the trend lines.
    b. Identify and label special points.
    c. Identify stratigraphic boundaries.
    d. Identify changes in T- and L-directions.

6. Construct cross section parallel to T-direction.
    a. Project the wellbore survey onto the T-section.
    b. Project average dip bars along the wellbore path.
    c. Plot special points at appropriate depths.
    d. Sketch structural form lines or use the isogon
        construction method to draw form lines.

7. Construct a contour map in the vicinity of the well.
    a. From the T-section select an horizon to map.
    b. Plot the well and T- and L-directions on the map.
    c. Plot L-direction features (inflection, trough,
        crestal, axial, and fault planes) on the map.
    d. Estimate the regional gradient and draw contours
        or use the plunge projection technique.

interpretation can be employed to derive results from the
dipmeter plots and construct cross sections and contour maps to
portray the geologic interpretation. The steps of a general
scheme is outlined (Table 1) and the use of relevant graphic
displays is exemplified.

The Ridge Circle Field of Fremont County, Wyoming (Levorsen, 1967) is an example of a well-defined geologic structure in the subsurface and reported in the literature. Hypothetical SCAT plots are used to demonstrate their use in interpretation of geologic structures. A straight vertical borehole is assumed in the example.

Borehole Survey Block Diagram

In offshore and tundra drilling locations, drill platforms or gravel pads are used for directional drilling of wells to establish a drilling pattern for efficient hydrocarbon recovery. The orientation measurements of azimuth, relative bearing, and deviation comprise a borehole survey which is used to calculate the dips relative to the horizon. This survey data also may be used to construct a block diagram displaying the borehole orientation in space. The borehole is drawn in perspective (Fig. 5) so that the well appears to come outward from the plane of the paper with increasing depth. In other words, the wellbore will be in a similar orientation for different wells and the orientation of north-south and east-west axes is selected automatically to show the best perspective.

The plot represents a partial cube with the two rear inside walls and the base drawn with a grid. The perpendicular projection of the borehole is shown on the north-south and east-west vertical sections of the cube. The well deviation in plan view is plotted on the base of the cube.

The borehole deviation survey is necessary in determining in which direction to project the well path onto a vertical cross sections. The relative orientation of the borehole and the geologic structure is critical to interpretation. Similar patterns on SCAT plots can be generated by a deviated borehole crossing structure or a structure with more than one phase of deformation could be penetrated by a straight well; for example, a diagonal axis could cross an anticlinal axis or a curved anticlinal axis could be penetrated by a straight hole.

Vertical Zonation

Determination of how many vertical zones may exist in a study well is achieved by examination of a continuous azimuth vector plot in conjunction with tadpole displays or the azimuth versus

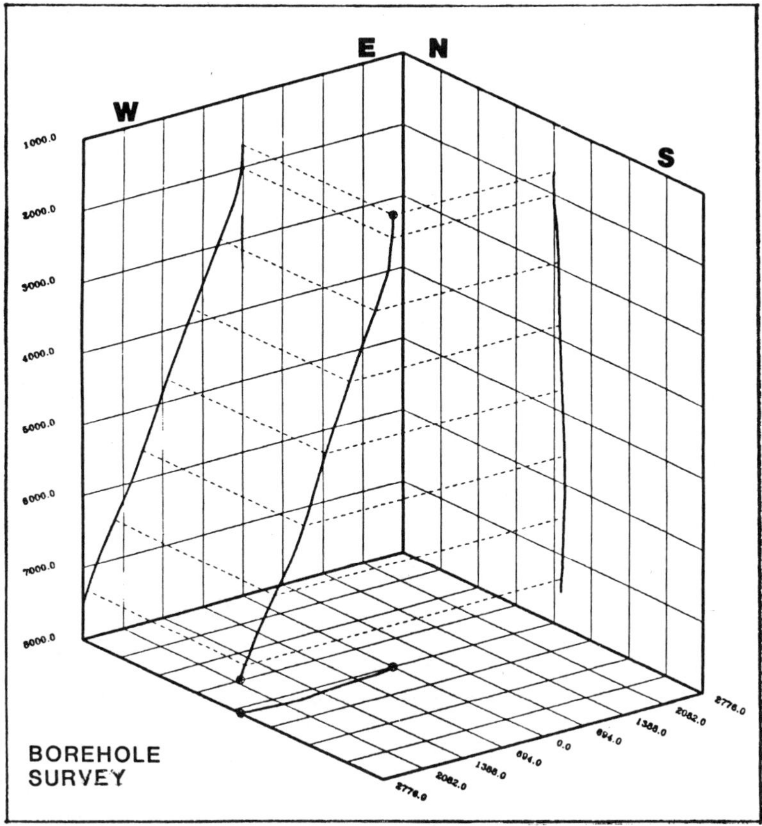

Figure 5. Block diagram display borehole orientation generated by INCLINE. Direction of drilling is close to due west.

depth plot (A-plot). The continuous azimuth vector plot was designed by Shields and is a graphic option in the INCLINE program. Only the direction of dip is used. Beginning at the bottom of the log and working upward, an arrow is drawn as on a map to show the azimuth of the first dip. The azimuth of the next dip is shown by another arrow with its tail connected to the head of the previous one. The resulting plot (Fig. 6) is a string of arrows with the depth written periodically along the path. The maplike plot may be drawn at any scale over any interval of the log. There may be so many dips in a section that the arrows are nearly invisible to the naked eye; the depths indicate which way the zigzag pattern proceeds. The vector plot is used to identify

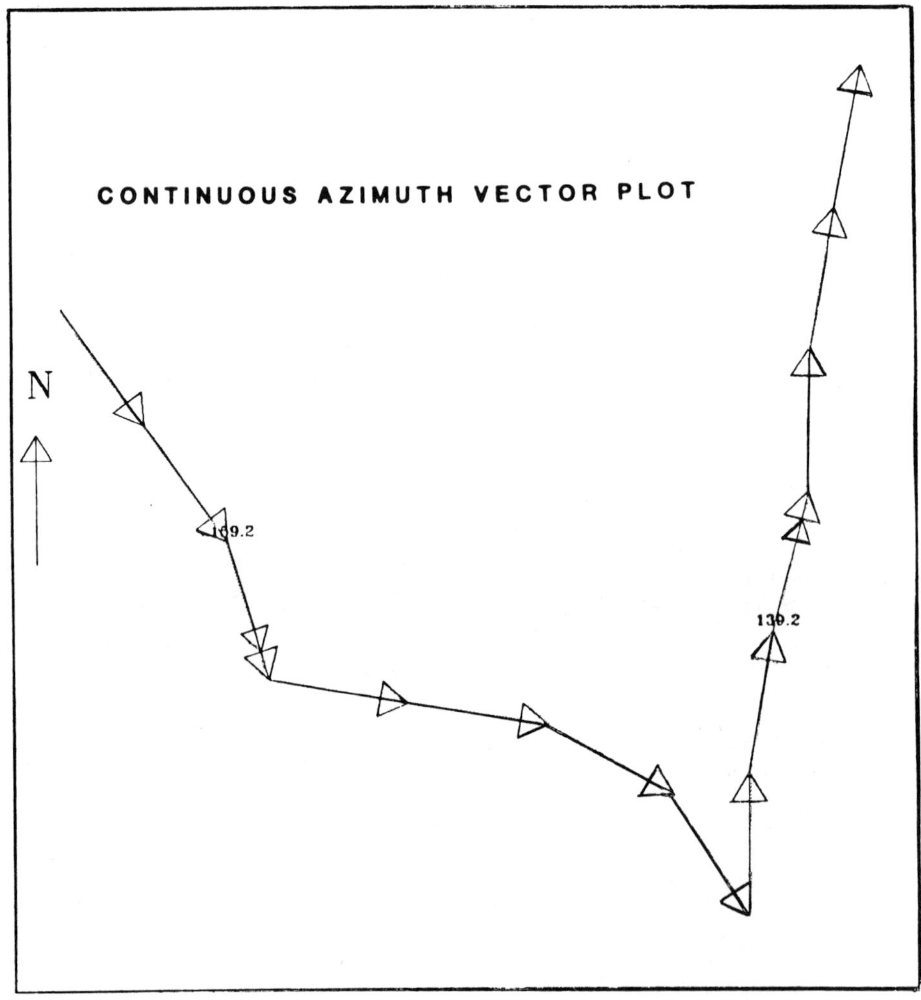

Figure 6. This continuous azimuth vector plot displays sudden
uphole change in azimuth from SE to NNE.

unconformities or detachments or other vertical boundaries
where the transverse and longitudinal directions may change.
The azimuths may be scattered, but they generally display a linear
trend. Changes in trend may indicate a change in structural or
stratigraphic T- and L-directions. Vertical zones with different T-
and L-directions may be proceed independently to produce SCAT
plots for the appropriate dip component orientations.

Dip Versus Azimuth (DVA) Plot

The general structural setting can be ascertained by pattern
analysis on the dip versus azimuth (DVA) plot (Bengtson, 1980a,
1982).

Pattern	General Setting
Low dip, scattered azimuth	Zero dip
Low dip, converging azimuth	Low dip
Moderate dip, concentrated azimuth	Moderate dip
Railroad track	Nonplunging fold
Horseshoe	Plunging fold
Christmas tree	Doubly plunging fold

The regional dip and plunge can be estimated from the patterns,
on the DVA plot. Only a partial pattern may appear, perhaps only
one limb of a fold is penetrated by the well. However more than
one pattern maybe superimposed upon the plot. Faults tend to
produce a partial horseshoe pattern from the drag zone formed by
a small anticline on the upthrown block and a small syncline on
the downthrown block. This effect is only displayed for faults
with a drag zone. The T- and L-directions can be estimated from
the DVA for the diagnostic patterns that are identified.

A useful modification of the DVA plot is to plot a symbol for each
data point that represents the well. For instance, ten different
symbols may be used and the depths can be divided into ten
groups with each represented by a different symbol. With this
option the depth of partial and complete patterns and their parts
can be identified immediately.

The hypothetical DVA plot for Ridge Circle Field (Fig. 7) displays a
partial railroad track pattern originating from two limbs of a fold.
In both limbs the higher dips are not present because the fold
may be of low amplitude, it may have a nonvertical axial plane, or
perhaps it is asymmetrical. The T-direction of the fold is
northeast or southwest along the central azimuth of the "railroad
tracks", or dip trends of the two limbs. The L-direction of the
fold is northwest or southeast or perpendicular to the fold is
northwest or southeast or perpendicular to the T-direction. The
partial horseshoe pattern associated with the northeastern limb is
created by the drag zone of a fault that cuts the northeastern limb.
The horseshoe is formed by the vectoral addition of the northeast

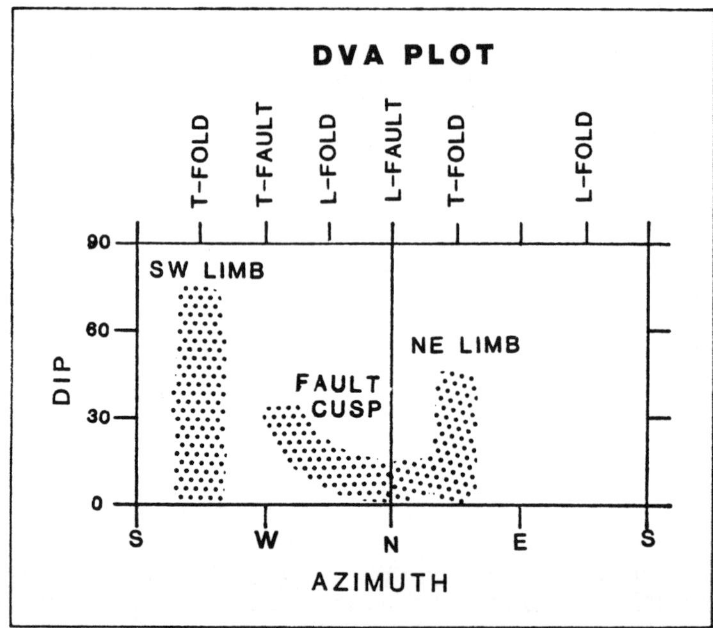

Figure 7. Hypothetical dip versus azimuth (DVA) plot for Ridge Circle Field.

dip of the fold limb and increasing dip to the west in the drag zone. The T-direction of the fault is to the west. Because the fault influences only a small vertical section of the well, the T-direction of the fold is appropriate for orienting the depth plots A short SCAT plot over the fault zone using its T-direction may be useful.

Depth Plots (A-, D-, T-, L-Plots)

The depth plots have increasing downward depth on the vertical axes (see Fig. 8). The horizontal scales are described as follows: azimuth (A-plot) has north (0 degrees) in the center with east (90 degrees) to the right and west (270 degrees) to the left, and south (180 degrees on the edges. Dip (D-plot) increases from left (0 degrees) to right (90 degrees). The Transverse (T-plot) dip components have zero degrees in the center and increasing to 90 degrees to both the right and left.

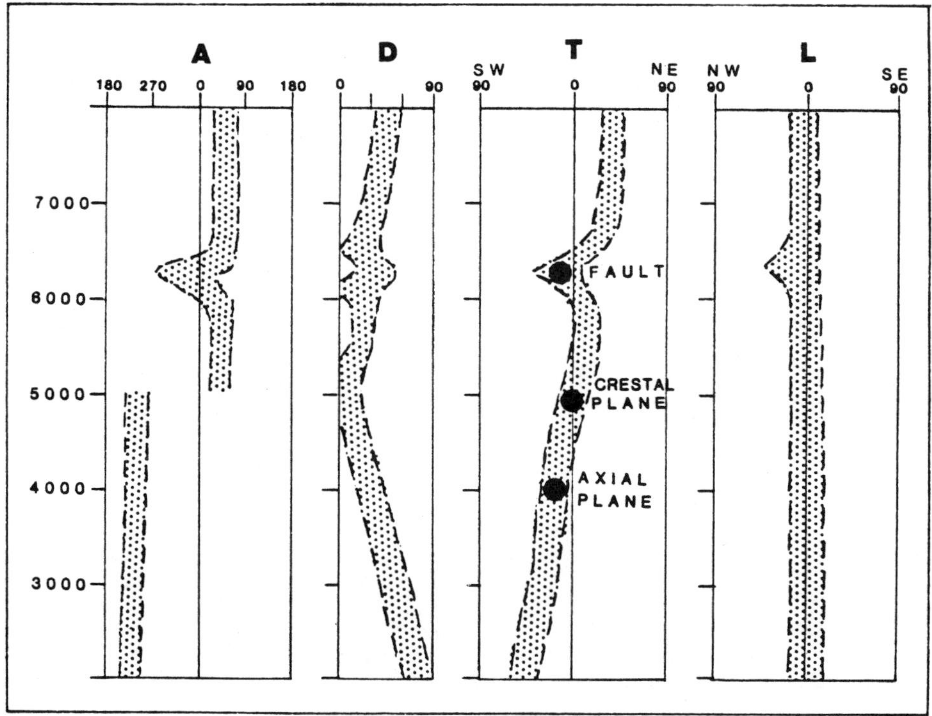

Figure 8. Hypothetical depth plots for Ridge Circle Field: Azimuth (A), Dip (D), Transverse (T), and Longitudinal (L) dip component.

Two parallel trend lines usually are drawn outlining each side of the dip concentration on the depth plots (Fig. 8). This practice is advocated by Bengtson and as simple as it seems, it is useful. The purpose of the lines is to give the mind a general pattern to analyze as well as to attempt to focus on the signal and de-emphasize the noise in the dip data.

The next step is to identify special points on the depth plots. These points coincide with the depth at which the dip trend responds to a structural element that has a geometrical definition or a graphic description. Special points include fault cusps; curvature boundaries of fault drag zones; crest, troughs, inflections, and axes of folds; detachment surfaces; plunge and symmetry reversals; and unconformities. Individual special points are characteristic on various SCAT plots, but the most important ones are those on the T-plot because they are used for cross-section construction.

Stratigraphic boundaries may be identified on T- and L-plots by an abrupt change in width of the trend or the width between the two trend lines. Generally the coarser the sediments, the greater the scatter in magnitude and azimuth of dip, and the greater the width of the trend of data points. Some stratigraphic boundaries also may be identified by a change in magnitude or character of the microresistivity curves on a log or plotted with an expanded depth scale.

A horizontal offset of the trend lines usually indicates the presence of an unconformity or detachment at the depth. The T- and L-directions may change orientation at such depths and reprocessing with refined T- and L-axes may be appropriate. A change in T- and L-directions is evident by changes in pattern on the T- and L-plots. Usually the L-plot displays a vertical band of data concentrated at zero for a nonplunging structure. If the T-plot shows no change in dip or plunge and the L-plot does, it is likely there is a change in T- and L-orientation. Changes in azimuth on the A-plot should verify those features identified on the dip component plots.

For the Circle Ridge Field the depth plots (Fig. 8) have some curvaceous trends. Except for one peak pointing to the northwest, the L-plot shows little change with depth indicating that the correct L-direction has been selected. Alignment along the zero axis indicates the geologic structures encountered have no plunge. The northwestern peak coincides with the depth of the fault suggesting that the T-direction for the fault is different than that for the fold.

The A-plot shows both limbs of the fold; the one with northeast dip is above the one with southwest dip. At the fault depth the azimuth is to the west indicating maximum dip in the drag is to the west. Because the dips above and below the fault are to the northeast, the fault must be within the northeastern limb.

The D-plot displays patterns in dip magnitude, but the T- and L-plots usually are more helpful because the dip components are oriented and right and left dip are distinguished. The fault is identified by the diagnostic cusp on the T-plot. Because the dip in the drag zone is to the west, the L-plot has a cusp also.

Where the trend crosses the zero dip axis of the T-plot, a trough or crest of a fold is penetrated. The point of minimum curvature on the T-plot is the axial plane of the fold.

Cross-Section Construction

The cross section is the most valuable tool to convey structural analysis to other geoscientists. The most useful cross section is drawn parallel to the T-direction, as though looking in the L-direction. The first step in cross-section preparation is to project the borehole onto T-section. For the rock units encountered to be at the proper depth on the cross section, the borehole path must be projected parallel to the L-direction and at the angle of plunge.

Once the path of the borehole is drawn, average dip bars are plotted at appropriate intervals. The transverse dip components are plotted for each value taken as an average over an interval of perhaps 50 or 100 feet (or meters) over stratigraphic intervals, if their thicknesses are appropriate. The special points are plotted at their respective measured depths along the wellbore. The structural interpretation is drawn by sketching form lines or using the isogon construction method (Ramsey, 1967). The patterns identified on the SCAT plots are most important in drawing an accurate cross section. Illustrating the major structural patterns first and adding the minor details later has proven more successful than the reverse procedure. The transverse cross section is emphasized here, but in some situations the longitudinal one may be significant also.

The special points are defined; and before a cross section can be drawn, some questions must be answered: Is the fold an anticline or syncline? Is the fault normal or reverse?

For folds with near vertical axes the crestal plane associated with an anticline will be penetrated before the axial plane; likewise, the axial plane of a syncline will be encountered before the trough plane. The identification of the intersection of the axial plane and the borehole maybe difficult. Theoretically it occurs at a point where the dip is increasing, but the rate of change of dip is changing from increasing to decreasing downward. In other words, the T-plot trend may have a slight flattening in the vicinity of the axial plant. This special point may be difficult to determine

with scatter in the data. An alternative is to draw two cross
sections with separate anticlinal and synclinal interpretations and
decide which is supported by other well data or seismic evidence.

The question of reverse or normal fault is a more specific situation
of the more general question of extensional versus compressional
tectonic setting. Additional information is necessary to decide
upon an answer. Repeat stratigraphic sections which occur on
conventional logs indicate the presence of reverse faulting in
compressional setting, unless a highly deviated well penetrates a
normal fault from the underside. Missing stratigraphic section
can be detected only with knowledge of the entire section; hence
more than one well usually is necessary to verify this situation.
Missing section may indicate normal faulting in an extensional
terrain unless a highly deviated hole penetrates the underside of a
reverse fault. The possibility of extensional features superimposed
upon compressional ones or vice versa must not be overlooked.

The transverse cross section for Circle Ridge Field (Fig. 9) is
constructed from the T-plot interpretation. The well path and

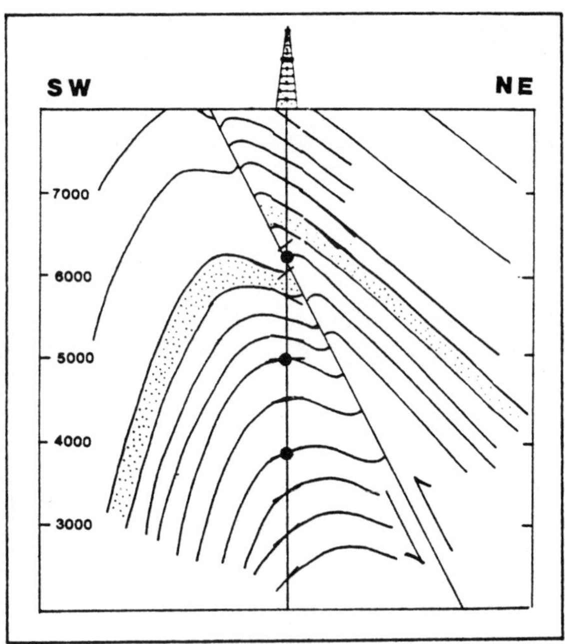

Figure 9. Cross section for Ridge Circle Field drawn from SCAT
plots.

dip bars are drawn first. The special points are plotted with their linear expressions drawn away from the wellbore as the axial, crestal, and fault planes. The axial plane is perpendicular to the dip at the special point. The crestal plane is parallel to the axial plane. The fault plane dips in the direction opposite the direction the cusp points on the T- and L-plots. The fault plane dips east in this situation.

Contour Mapping

The horizon to map is selected from the T-section. The deviation survey is used to determine location of the intersection of the well and the surface to be mapped. The T- and L-directions may be plotted. The regional gradient is sketched from the T-plot. It may be appropriate to draw L-direction features such as the trace of fault, inflection, crestal, trough, or axial planes.

The structural configuration is drawn by sketching in contours or by plunge projection techniques (Mackin, 1950; Regan, 1972). As is the situation with cross sections, the various SCAT plots provide converging lines of supporting evidence for a unique interpretation of geologic structures.

The hypothetical contour map for the Circle Ridge Field (Fig. 10) is constructed from the cross section and SCAT conclusions. The fold axis and fault are drawn relative to the borehole. Dip of the fold limbs is derived from vertical changes in the well. The mapped horizon overlaps itself along the fault zone.

Comparisons of this hypothetical dipmeter interpretation (Fig. 9 and 10) can be made with the field map and cross section (Fig. 11) from Levorsen (1967, p. 277). The short coming of the SCAT analysis is that the domal structure is not detected unless a plunge reversal is encountered in the well. The purpose of this hypothetical example is to demonstrate the methods and reasoning used in interpretation of the computer-generated plots.

Fracture Information:
Microresistivity Anomalies in Azimuth Space

An unique plot generated by the INCLINE program shows the microresistivity anomalies in azimuth space (Fig. 12). This graphic display in log format is designed to yield information about borehole eccentricity and the orientation of microresistivity variation. The left track of the log displays the difference in the

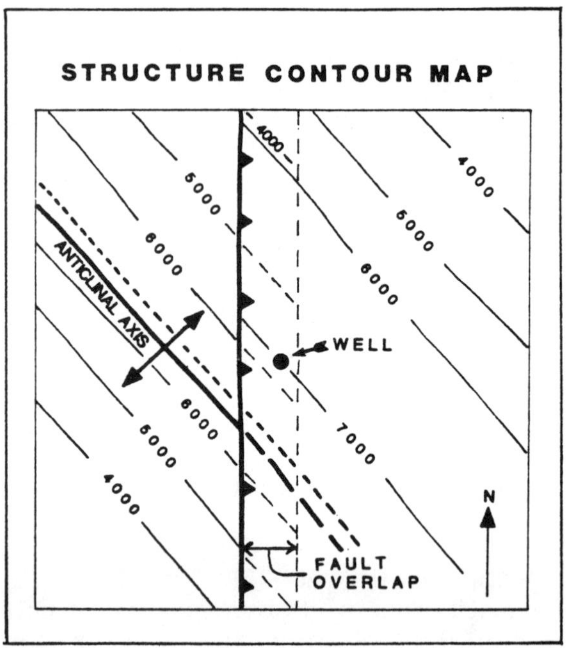

Figure 10.   Structure contour map on top of speckled unit on
             section.

two perpendicular sets of calipers by the width of the curve and
within the track the location of the left edge of the curve
indicates the orientation (0 to 360 degrees) of either the first or
second arm. Because there are two sets of calipers, there are two
curves displaying the difference in borehole diameter in their
respective orientations.

The four microresistivity curves also are plotted in azimuth space
(0 to 360 degrees) in the second track. The curves are the
difference between the individual measurements for a pad on an
arm and the average of all four. Zones of anomalous resistivity may
indicate the presence and orientation of fractures. If indeed,
fractures have vertical anomalies, their orientation can be
determined in situations of tool rotation. As each arm rotates past
the anomalous azimuth, the microresistivity anomaly will reach its
maximum. The caliper differences are used in conjunction with
microresistivity data to verify borehole ellipticity in the direction
of fracture planes, if they are open fractures. In deviated wells the

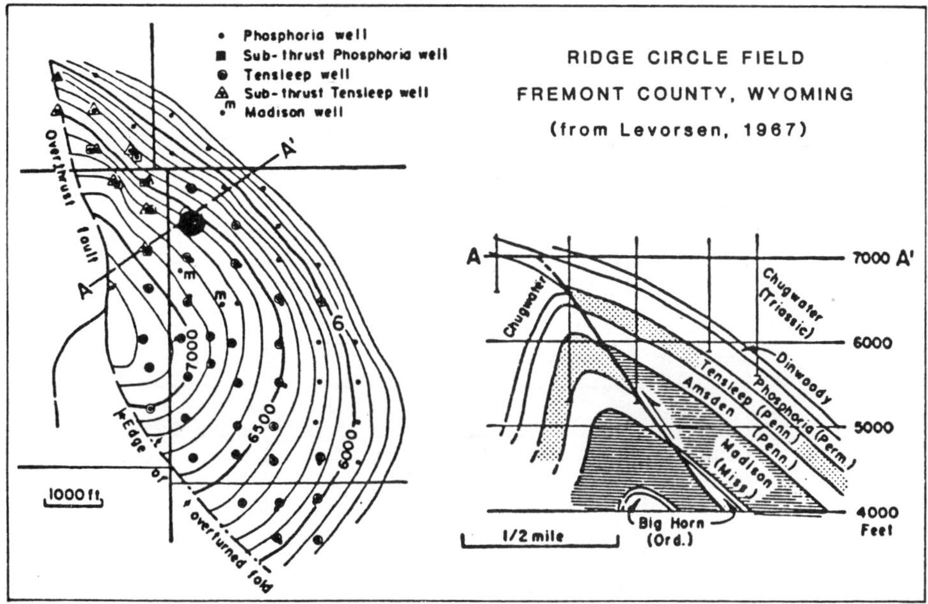

Figure 11.   Hypothetical well used to illustrate SCAT method of
             dipmeter interpretation is shown at midpoint of A-A'.

borehole may be elliptic with the long axis parallel to the
direction of drilling. This drilling effect must be considered
when attempting fracture identification. Fracture information
may be incorporated with the structural model developed by SCAT
usage or employed in reservoir development.

CONCLUSIONS AND FUTURE ADVANCEMENTS

The trend to increase the quantity of measurements and improve
their geometric relationship is established in current
developments. Computer programs to process the data will be
capable of finer resolution of vertical resistivity contrasts which
may provide more stratigraphic and sedimentologic information,
but the limitations remain with scatter in dips and limited
sampling around the borehole circumference. Some of the
filtering techniques of spectral analysis employed in seismic
processing may be capable of noise reduction to reduce data
scatter. Vector averaging is another technique that has been used

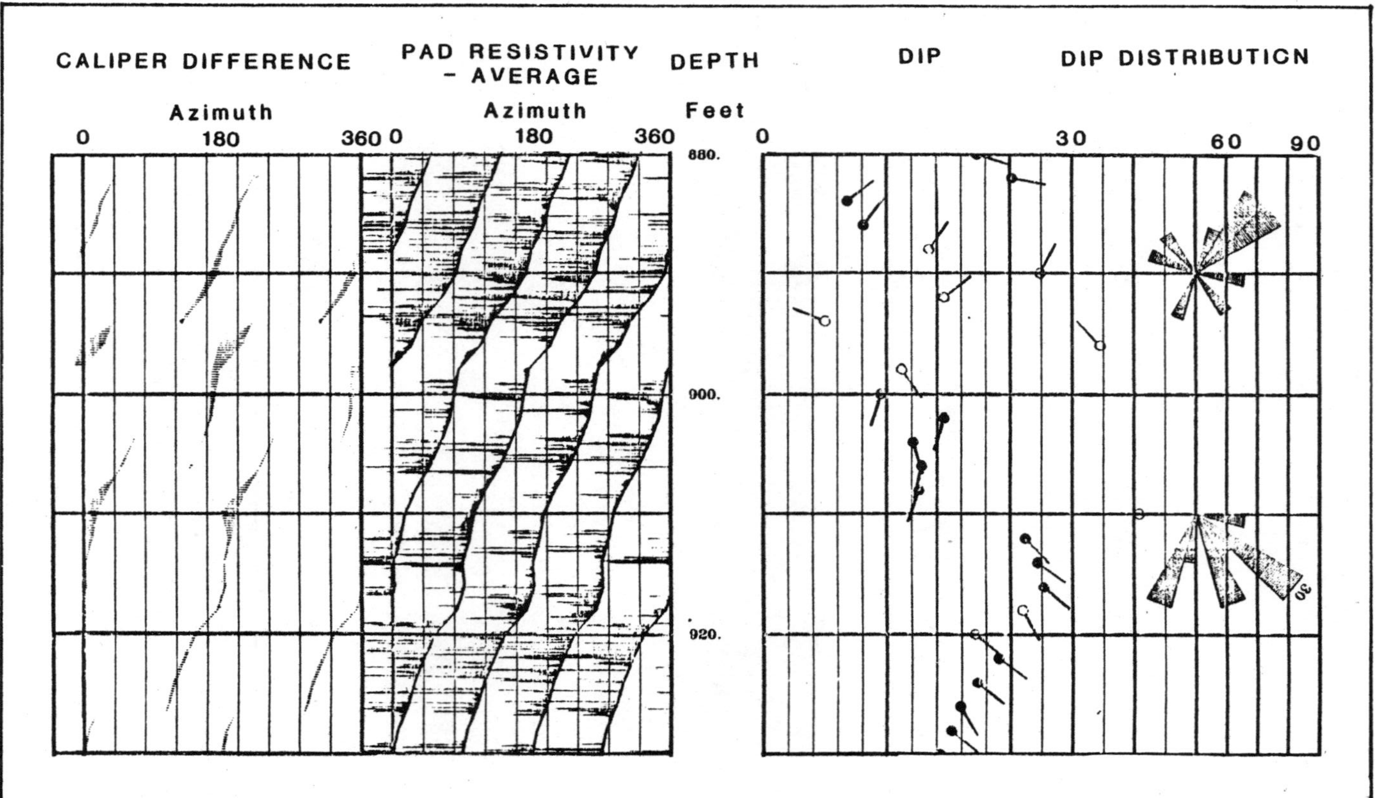

Figure 12. Microresistivity anomalies in azimuth space, a log format generated by INCLINE.

to reduce scatter in magnitude and azimuth of dips. The sampling distribution with respect to the borehole may be improved by televideo recording or some type of radially scanning measurement system. The capability of reproducing the cylindrical surface of the borehole as a graphic image is already underway. Experimentation with measurement of different physical properties is likely.

Once the digital data from dipmeter logs in used widely in petroleum companies, more software for interpretation will be developed. The trend in geologic interpretation may follow that of seismic processing in the use of modeling. In modeling, artificial intelligence programs can be developed to identify the dip trends and display the results on cross sections and maps with some input from the geologist. This method, in which dipmeter data is used to construct a model of geologic structure, is inverse modeling. Forward modeling also can be employed whereby a set of geologic and geometric descriptors can be specified and a computer program can simulate the dip results. Reiterative changes to the input parameters can be made until the simulated results match the original data. The foundation for both types of computer analysis is well established in Bengtson's SCAT.

The future looks promising for dipmeter advancement. Many exploration wells have been logged in the past and many have the potential for improved processing and interpretation. Dip meter logging in the future will have the advantages of improved instrumentation, greater sampling capacities, computer processing with a geologic basis, and more accurate tools for interpretation.

REFERENCES

Bengtson, C.A., 1980a, Statistical curvature analysis methods for interpretation of dipmeter data: The Oil and Gas Journal, v. 78, no. 25, p. 172-190.

Bengtson, C.A., 1980b, Structural uses of tangent diagrams: Geology, v. 8, no. 12, p. 599-602.

Bengtson, C.A., 1981, Statistical curvature analysis techniques for structural interpretation of dipmeter data: Am. Assoc. Petroleum Geologists Bull., v. 65, no. 2, p. 312-332.

Bengtson, C.A., 1982, Structural and stratigraphic uses of dip profiles: Am. Assoc. Petroleum Geologists Mem. 32, p. 31-45.

Bengtson, C.A., 1983, Easy solutions to stereonet rotation problems: Am. Assoc. Petroleum Geologists Bull., v. 67, no. 4, p. 706-713.

Levorsen, A.I., 1967, Geology of petroleum: W.H. Freeman and Co., San Francisco, 724 p.

Mackin, J.H., 1950, The down-structure method of viewing geologic maps: Jour. Geology, v. 58, no. 1, p. 55-72.

Ramsey, J.G., 1967, Folding and fracturing of rocks: McGraw-Hill Book Co., New York, 568 p.

Regan, D.M., 1973, Structural geology: John Wiley & Sons, New York, 208 p.

Rider, K.L., 1978, Dipmeter log analysis: an essay: Soc. Prof. Well Log Anal. Symposium, v. 19, p. 1-18.

CRUSTAL ABUNDANCE MODELING OF MINERAL RESOURCES: RECENT INVESTIGATIONS AND PRELIMINARY RESULTS

DeVerle P. Harris

University of Arizona

ABSTRACT

This paper reviews the evolution of crustal abundance models from the early univariate stock-based models of element concentration, to a bivariate statistical model of deposit size and grade, and ultimately to a multivariate model (4-dimensional). The extension to a multivariate statistical model is motivated by the desire to use a crustal abundance model to estimate resources and potential supply. Such use requires that the model possess the structure to permit credible costing of exploration and production. Accommodating such a requirement leads to specification of a statistical model that describes deposit size, grade, depth, and intradeposit grade variance. This paper reports on a preliminary investigation of the estimation of such a model for U. S. uranium.

INTRODUCTION

Perspective

Crustal abundance, the ultimate stock of an element or compound, has been used as a foundation for making inference about the quantity and quality of mineral occurrences of a higher order of concentration, such as those occurrences that support mineral industries. The subject matter of this paper is the use of crustal abundance to estimate and describe mineral resources and potential mineral supply. Clearly, use requires an interest in long-term mineral availability or resource adequacy. Such interest is assumed in the writing of this paper.

Terminology

The term mineral will be used to represent minerals and mineraloids. Mineral endowment refers to the quantity of the mineral that is present within the region, given cutoff values for deposit size and grade and depth to deposit. Mineral resource refers to the quantity of the element or compound that could be produced economically, given current or near feasible technology and given a specified price. Potential mineral supply is the amount that is both producible and discoverable for a specified price and given current exploration efficiencies. For an in-depth development of mineral resource concepts and terms see Harris (1984a, 1984b).

Motivations for Crustal Abundance Modeling

The appeal of crustal abundance as a foundation for mineral resources estimation seems to be due to the following features, some of which may be relevant only in a limited sense:

> Mineral resources are linked to a basic and fundamental physical feature of the Earth's crust - the concentration of the element (compound) of interest.

> Crustal abundance methods of appraisal are quantitative, as opposed to the traditional subjective evaluations by geologists .

> Crustal abundance aggregates beyond the troublesome details of geologic information, thereby avoiding the need for explicit geological analysis.

> Because of aggregation beyond geological particulars, crustal abundance models permit (allow) modes (types) of deposit occurrence not recognized currently in the region.

Given on the one hand that the abundance of an element is simply the product of its average concentration and the weight of the crust, and on the other hand that here our interest in crustal abundance is to use it to estimate mineral resources and potential supply, one or more models of physical or economic relations must be employed to bridge from the simple physical concept of resource base to the compound physical-economic concepts of mineral resources and potential supply. Clearly, strong assumptions are necessary, for example grade continuity. These may be challenged on technical grounds (Harris and Skinner, 1982; Harris, 1984a; Skinner,1976). Furthermore, it is here acknowledged that some of the attractive features of crustal abundance models, such as a mathematical relationship that can be manipulated easily to examine the consequence of low grades, can be obtained only at a cost. This cost is the nonuse of other disaggregated and specific geologic information - see Harris (1984a,1984b) for a more complete examination of these issues. Here, let us suppose that the benefits of aggregation and generalization outweigh costs and, therefore, limit the subject matter to identification and estimation of a comprehensive crustal abundance model for the estimation of resources and potential supply.

Structure of Crustal Abundance Models

The task of estimating mineral resources and potential supply from crustal abundance is complex and belies the simplicity of the crustal abundance concept. Complexity arises from (1) the use of limited and diverse information on known deposits to identify and estimate a mathematical model that bridges from crustal abundance, the foundation, to the stock of mineral and ore deposits that comprise the resource base, and (2) the consideration of economics in the transformation of unknown endowment to reserves and potential supply.

Three main lines of inquiry to crustal abundance modeling have emerged:

>the relationship across elements of crustal concentration to ore reserves (McKelvey, 1960; Erickson, 1973) or to ore reserves plus cumulative production (DeNapoli, 1976).

>quantity-quality (tonnage-grade) relationship for a single element or compound (Singer and DeYoung, 1980; Harris and Agterberg, 1981).

the probability (geostatistical) modeling of the "parent" distribution of element (compound) concentration (Brinck, 1967; Agterberg and Divi, 1978; Deffeyes and MacGregor, 1980; Harris and Chavez, 1981).

Scope of Paper

This paper pursues the last of the outlined approaches and can be seen as an extension of previous work, emphasizing a systems approach to the <u>use of information on ore reserves, production, ore-deposit characteristics, the exploration process, and development and production costs</u> to estimate the parameters of a more complex crustal abundance model, with U.S. uranium serving as a case study. However, because some of the properties of this more comprehensive model result from the identification of one or more problems or inadequacies of earlier models, a brief review of some developments and practices is presented next.

SOME IMPORTANT EARLY DEVELOPMENTS AND IDEAS

Brinck's Initial Model

Brinck (1967), drawing upon the classic papers of Ahrens (1953, 1954) on the lognormal distribution of element concentration, McKelvey's (1960) examination of reserves and crustal abundances, and the variance-volume relations of DeWijs (1951, 1953) and Matheron (1971), was the first to base an estimate of mineral resources upon a probability model of element concentration. His approach in its most straightforward application was demonstrated on Oslo, Norway, for which he had metal concentrations in samples obtained from a geochemical survey. The parameters of the normal distribution of the logarithms of element concentrations were estimated directly from moments computed on the geochemical samples. Brinck then modified this distribution, using the Matheron-DeWijs variance-volume relation, to one that described the distribution of concentration (grade) in "typical" deposit-size accumulations. The quantity of mineralized rock (R) comprising resources of the element within the region was estimated by multiplying the fraction of all deposits of the typical size that have grades above a specified cutoff grade, q' and W, the weight of the crustal material to the specified depth:

$$R = W \int_{\ln q'}^{\infty} f(\ln q; \mu, \sigma^2)\, d\ln q \, ,$$

where $f(\ln q; \mu, \sigma^2)$ is the probability density function (pdf) of grade in "typical" deposit-size accumulations

μ, σ^2 are the parameters of the lognormal pdf

μ as defined by Brinck, is $\ln(A)$, where A is crustal concentration.

Brinck described a second approach to estimation of the parameters of f when no geochemical data are available. This approach requires the identification of "typical deposit size" of the known inventory (known deposits). By making the assumption that all deposits having grades above the current mining grade (q') have been discovered, the ratio of $w_{q'}/W$ can be interpreted to lie on the pdf of crustal grades:

$$\frac{w_{q'}}{W} = \int_{\ln q'}^{\infty} f(\ln q; \mu, \sigma^2)\, d\ln q = 1 - F\left(\frac{\ln q' - \mu}{\sigma}\right)$$

where μ is assumed known (or reasonably well estimated) and equal to the logarithm of crustal concentration: $\mu = \ln A$

$$F\left(\frac{\ln q' - \mu}{\sigma}\right) = P(\ln Q < \ln q')$$

Because μ is assumed known, σ is calculated directly:

$$\sigma = \frac{\ln q' - \ln A}{z'}$$

Agterberg and Divi's Modification

The theme of this early work by Brinck was employed by
Agterberg and Divi (1978) for the distributions of crustal material
by grades of copper, lead, and zinc in the Canadian Appalachian
Region - see also Agterberg (1980). Unlike Brinck, however,
Agterberg and Divi interpreted the available measures of crustal
concentration, A, to be the <u>mean</u> concentration of a lognormal
distribution having parameters μ and σ^2 :

$$\ln Q \sim N(\mu, \sigma^2)$$

$$A = e^{\mu + \sigma^2/2}$$

Therefore, $\mu = \ln A - \sigma^2 / 2$, whereas Brinck equated μ to $\ln A$.

Employing w_q, for each of eight cutoff grades, Agterberg and Divi
(1978) made eight estimates of σ, one for each grade:

$$\frac{w_{q'}}{W} = p\,(\ln Q \geq \ln q') \implies Z$$

and

$$\sigma = z - (z^2 + 2A - 2\ln(q'))^{1/2} \ , \ z > \sigma$$

$$\sigma = z + (z^2 + 2A - 2\ln(q'))^{1/2} \ , \ z \leq \sigma$$

Determining estimated σ's to be similar for the four highest cutoff
grades but significantly smaller for the lower ones, the average of
the estimates for the four high cutoff grades was taken as the
estimate of σ, reasoning that estimates for the lower grades are
affected (biased) by incomplete exploration.

An Asymptotic Variance

Harris and Chavez (1984), using the 1978 preproduction
inventory of \$50.00 reserves plus production reported by the U.S.
Department of Energy (DOE), extended the approach
demonstrated by Agterberg and Divi (1978) of estimating σ for
each of several cutoff grades to the use of these multiple estimates
to estimate an asymptotic value of $\hat{\sigma}$ that is the value to which $\hat{\sigma}$,
tends as cutoff grade increases. They reasoned that exploration is
not complete really for any grade; consequently, even an average
of $\hat{\sigma}$'s for high grades is biased negatively. Furthermore, if
exploration were progressively less complete for successively
lower cutoff grades (q'), $w_{q'}$ also would be progressively less than
it should be as q' is decreased. Such an effect in $w_{q'}$ would
produce estimates of σ that would be successively smaller as q' is
decreased. Because of thorough data acquisition and reporting by
DOE, data were available on the magnitude of the preproduction
inventory for many cutoff grades, permitting the testing of this
hypothesis.

Figure 1 shows the estimated standard deviations associated with
each cutoff grade. This figure suggests that as the value of the cutoff grade increases, the associated estimate of σ approaches
asymptotically the true value of σ. A mathematical curve that
exhibits a pattern similar to that of the estimates of σ is the
modified exponential: $\hat{\sigma} = \sigma^* + ab^{q'}$, $a < 0, 0 < b < 1$. Thus, as
q' approaches infinity, $\hat{\sigma}$ approaches σ^*. Figure 1 shows the
modified exponential fitted to the standard deviation estimates.
The asymptotic value of the equation of this fitted function is
1.437, a value slightly larger than 1.3561, which was obtained by
averaging the standard deviations for the high cut-off grades.

Estimates of U_3O_8 endowment using models based upon an
averaging of selected estimates of σ and an asymptotic $\hat{\sigma}$ are
plotted in Figure 2, as also is the DOE inventory of U_3O_8. Figure
2 shows that the endowment model based upon the asymptotic
variance has the desirable property that for all cut-off grades,
estimated U_3O_8 exceeds the known inventory. Inasmuch as
exploration for uranium in the U.S. is far from exhaustive, this is a

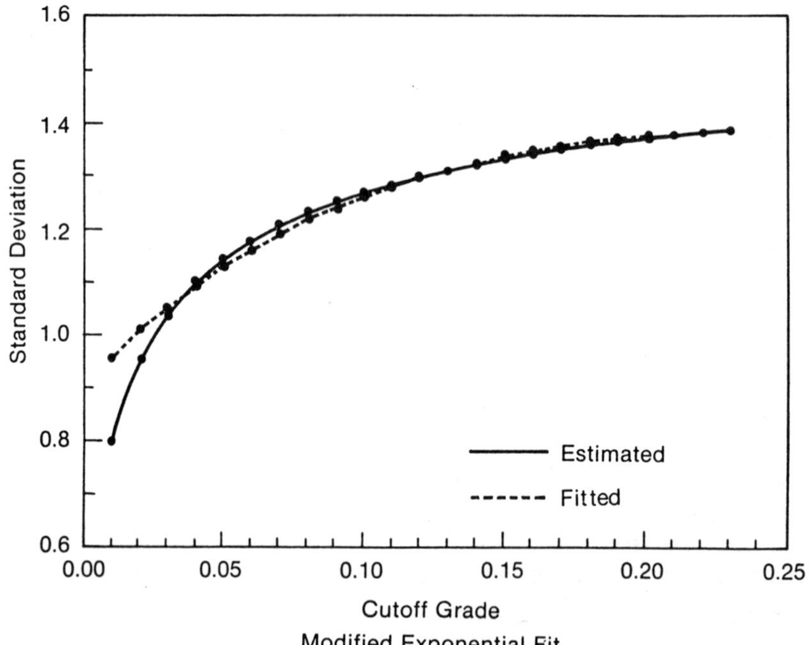

Figure 1. Modified exponential curve fitted to estimated standard deviations of U.S. uranium inventory grade categories (Source: Harris and others, 1981).

desirable result. Estimates of $U_3 O_8$ endowment by the model based upon the average of estimates of σ for the high cut-off grades does not possess this property. Endowment estimates for cut-off grades in excess of 0.175 are less than the known inventory, which is a disturbing result. By this criteria of evaluation, the model based upon the asymptotic $\hat{\sigma}$ is superior to that based upon the average of selected estimates of σ.

It is noteworthy that the crustal abundance model for copper that was estimated by Agterberg and Divi also underestimated endowment for the high grades, adding weight to the suggestion that the assumption made by Brinck that exploration has located all of the high-grade deposits leads to negatively biased estimate of σ.

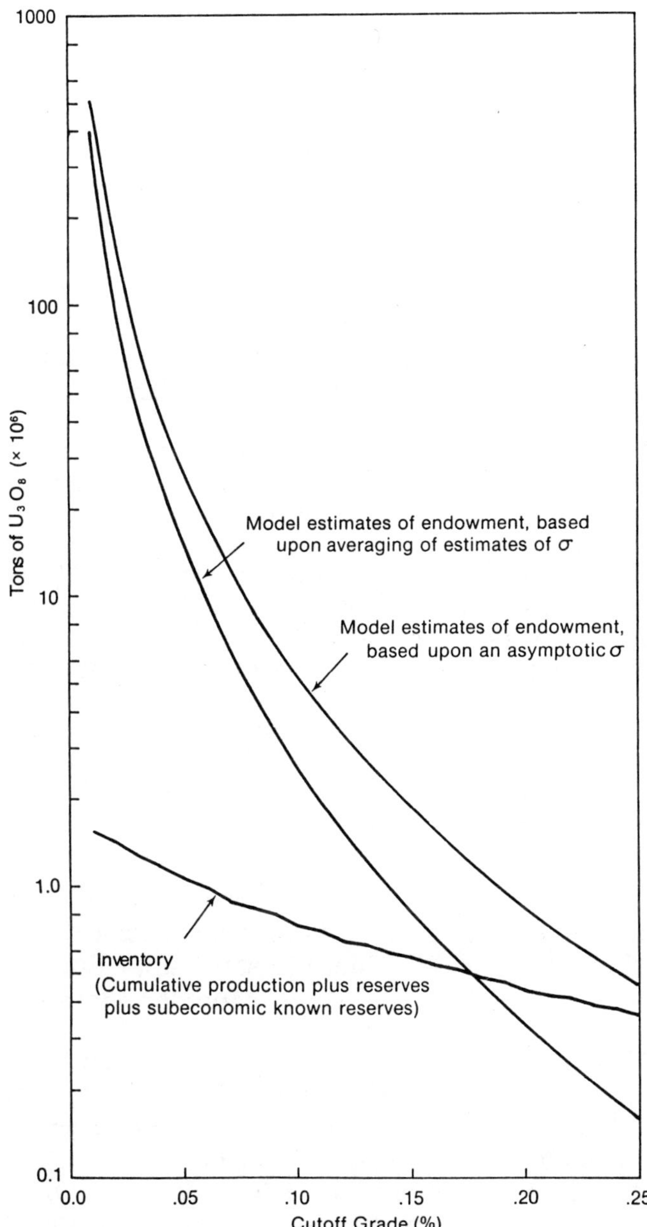

Figure 2. Comparison of estimates by lognormal models of uranium endowment inventory (Source: Harris, 1984a).

SOME COMPLICATIONS

Implication of Bias in the Variance Estimate

Determining a negative bias in the estimate of σ invites further critical examination of the ratio $w_{q'}/W$ as a probability of the crustal abundance model, a procedure demonstrated by Brinck (1967). The suggestion was made in the previous section that the negative bias results from considering $w_{q'}$ to be complete. It is noteworthy that $w_{q'}$ is only the numerator of a ratio and if incomplete data on $w_{q'}$ can lead to bias in an estimate of σ so also can an inappropriate specification of W, which is a function of area (A), depth (h), and weight factor (f): $W = A \times h/f$.

A source of difficulty in this relation is the appropriate value for h. Consider, for example, U.S. uranium - most of the U.S. deposits that constituted the 1978 preproduction $50 inventory occur at depths no greater than about 600 ft., but a few are at depths in excess of 2000 ft. What is the appropriate value for h: 600, 2000, or something in between? It may be argued that depth selection makes no difference as long as the inventory also is adjusted, indicating those deposits of greater depth are removed from the inventory. Although this adjustment is appropriate, it is not sufficient really if depth is an important determinant of exploration cost and performance, or of mining cost, for deposits of the deeper levels would be less well represented in the inventory than those of shallow levels, indicating that $w_{q'}$ is incomplete even after restricting W by depth and adjusting $w_{q'}$ to exclude those deposits having depths greater than the restricting depth. What seems to be needed is a method for estimating σ without employing $w_{q'}$ or W. There is some appeal to a statistical approach as an alternative, one which considers data on deposit size and grade to be <u>sample</u> information from which the required parameters can be estimated. This statistical approach avoids assumptions regarding $w_{q'}$ and obviates W in the estimation of parameters. A statistical approach was described and demonstrated by Drew (1977), but for a more extensive model. Before describing Drew's approach, it is useful to examine motivations for a more highly structured crustal abundance model.

Need for Modeling Deposit Tonnage

A crustal abundance model that consists solely of a probability density function of element concentration and the weight of the

Earth's crust offers limited provisions for the description of mineral resources or potential mineral supply, each of which has an economic dimension. Even the most rudimentary cost estimation requires deposit size (t) and grade (q), for these are major cost determinants: t is the source of economies of scale in mining, and q is the major factor in milling and processing design and cost. Furthermore man's experience is that a mineral occurrence in nature exhibits and is defined by <u>both</u> size and grade, and that both of these show great variation in nature. Consequently, a population of deposits is described properly by a joint (bivariate) distribution of t and q.

A BIVARIATE CRUSTAL ABUNDANCE MODEL

Extension by Drew

Drew (1977), striving to improve the usefulness of crustal abundance models in describing uranium resources and in the examination of the potential to meet future uranium demands, made the following extensions in crustal abundance modeling:

Replaced the univariate lognormal model of concentration (grade) by a bivariate model of deposit size and grade.

Obviated the use of $w_q{'}$, the weight of the inventory, and W, the weight of the crust in estimating unknown parameters in the bivariate model by adopting a more conventional statistical approach in which statistical data on deposit size represent a biased and truncated sample.

Identified (1) the need for a biased bivariate distribution to describe discoverable deposits and (2) the relationship between unbiased and biased distributions.

Identified a relationship of cost to deposit size and grade: this relationship serves to truncate the population of discoverable deposits, excluding those having sizes and grades so low that production cost is higher than price. Suppose that the joint pdf for ln t and ln q is a bivariate independent normal, having parameters μ_q, $\sigma_q{}^2$, μ_t, $\sigma_t{}^2$:

$$f\left(\ln t,\ \ln q;\ \mu_q\ ,\ \sigma_q^2\ ,\ \mu_t\ ,\ \sigma_t^2\right)$$

Then, given (1) a discovery relationship in which probability of discovery is proportional to t and q and (2) the properties of the normal distribution, f ' ( ), the biased distribution of discoverable deposits, is defined as follows:

$$f' (\ln t, \ln q; \mu_q', \sigma_q^2, \mu_t', \sigma_t^2) ,$$

where

$$\mu_q' = \mu_q + \sigma_q^2$$

$$\mu_t' = \mu_t + \sigma_t^2$$

$$\ln A = \mu_q + \sigma_q^2 / 2$$

Define c, mining and milling cost per lb $U_3 O_8$, to be a function of t and q:

$$c = Kt^{\alpha_1} q^{\alpha_2} = \lambda(t,q)$$

Then,

$$\ln c = \ln K + \alpha_1 \ln t + \alpha_2 \ln q$$

Or, generally

$$\ln c = \phi (\ln t, \ln q)$$

Consider a maximum cost, which is equal to price, p. Then, from the function ϕ, two inverse functions can be defined:

$$\max c = p$$

$$\ln q = \phi_q^{-1} (\ln t, \ln p)$$

$$\ln t = \phi_t^{-1} (\ln p, \ln q)$$

Suppose that from available deposit data, the means $\bar{t}$, $\bar{q}$,

and $\bar{m}$ (m = t x q) can be computed. By appropriately employing
$f'(\ )$, $\phi_t^{-1}(\ )$, and $\phi_q^{-1}(\ )$, mathematical expectations equivalent in
concept to these statistics can be defined, giving three equations
in three unknowns - although the independent bivariate normal
model requires four parameters, only three here need to be
estimated, for two of the four are related through the crustal
concentration (A): $\ln A = \mu_q + \sigma_q^2 / 2$ and permitting a solution
by nonlinear methods. As an example of these three equations,

consider the one for $\bar{q}$:

$$\bar{q} = \frac{\displaystyle\int_{-\infty}^{\infty} \int_{\phi_q^{-1}(\ln t,\ln P)}^{\infty} q\, f'(\ln t,\ \ln q;\ \mu_q',\ \sigma_q^2,\ \mu_t',\ \sigma_t^2)\ d\ln q\ d\ln t}{\displaystyle\int_{-\infty}^{\infty} \int_{\phi_q^{-1}(\ln t,\ln P)}^{\infty} f'(\ln t,\ \ln q;\ \mu_q',\ \sigma_q^2,\ \mu_t',\ \sigma_t^2)\ d\ln q\ d\ln t}$$

Drew (1977), employing the following data and relations,
demonstrated his model on U.S. uranium deposits:

$$A = 2\ \text{PPM}\ (U_3 O_8\ \text{equivalent})$$

$$\bar{t} = 100{,}000\ \text{mt}$$

$$\bar{q} = 2600\ \text{PPM}$$

$$\bar{m} = 190\ \text{mt}\ U_3 O_8$$

$$p = \$8\ /\ \text{lb}\ U_3 O_8\ (1\ \text{Jan. 1970})$$

$$c = 0.66\, t^{-0.159}\, q^{-1.0}$$

These inputs yielded the following parameters of the full (unbiased and untruncated) bivariate population:

$$\hat{\mu}_q = -15.10$$

$$\hat{\mu}_t = 7.517$$

$$\hat{\sigma}_q^2 = 3.96$$

$$\hat{\sigma}_t^2 = 1.362$$

Of course, $A = 0.000002 = e^{-15.10 + 3.96/2.0}$. These parameters later will be compared to recent estimates from an even more highly structured crustal abundance model, but first it is useful to examine what is learned from Drew's model.

First, replacing the use of $w_q{'}$ for parameter estimation by a statistical procedure incurs the cost of considerably more structure in the crustal abundance model. Much of this structure reflects the introduction of exploration performance and mining cost. Because statistical data on deposit size and grade exist primarily because of economic activities, introducing economics is unavoidable; it is present in exploration experience, and its influence is present in basic deposit data. This concept will be developed further, but first, let us examine a fundamental statistical issue, dependency between deposit grade and tonnage.

The Issue of Correlation

As in research in general, efforts to resolve one problem or issue identifies or highlights a new one. Drew's extension to a bivariate model is no exception to the rule. In this situation, the new problem (issue) is correlation between deposit size and grade. Does it exist? If so, how should it be modeled and estimated?

The assumption made by Drew (1977) of statistical independence seems reasonable as a first approximation, for it is supported by statistical studies of some types of deposits - (see Singer, Cox, and Drew, 1975; Harris, 1984a). However, independence of tonnage and grade in a crustal abundance model, which by definition represents all modes of occurrence of an element or compound, may be far less innocuous and less justifiable than it seems on the surface. Simple logic and the persuasion of experience cast doubt on the existence of independence through the entire range of element concentration.

For expository purposes, consider the element copper suppose that we define the unit for observation of grade to be the average size of mineral crystals of the most typical ore mineral of copper. Then, if in a large region, like the U.S., the Earth's crust to a depth of 1 mile were divided into samples of rock of this size and a relative frequency distribution of element concentration were constructed, there would be an exceedingly small, but nonzero, relative frequency for 99 to 100 percent concentration. Man's experience includes such an observation. But, if this region were divided into samples of rock of 100,000,000 tons, on the basis of man's experience, the relative frequency of 99 to 100 percent concentration would be zero; we have no evidence that such an event has ever occurred. Similarly, for grain or crystal-sized rock samples, reason allows that there could be a small, but nonzero relative frequency for a very low concentration, say 0 to 10^{-10} percent. But, for rock samples of 100,000,000 tons, we have no experience of such a low concentration. Is it really credible when contemplating all modes of occurrence of an element that size and grade can be independent statistically for all grades? Is it not possible that dependency between size and grade exist across modes of occurrence even though within some deposit types such dependency is not supported by statistical analysis (Harris, 1984a)? Finally, given contamination of data (sample) by exploration performance and production economics, how credible are previous statistical investigations of dependency that did not take into account economic truncation and translation?

ADDITIONAL MODIFICATIONS AND EXTENSIONS

Perspective

Modifications or extensions identified in this section result from examination of available deposit data and careful consideration of the ways in which the data depart from a random sample of a population of mineral occurrences. These modifications or extensions are additional in the sense that those identified earlier, although appropriate, are not enough, for one of the strongest determinants of discovery and of production costs is depth. However, tonnage, grade and depth are not enough, for grade differs within the deposit, and mining employs this variation to define the optimum cutoff grade and the associated quantity of ore and its average grade. These concepts are discussed further in the following sections.

Depth

Given that (1) the model is to be used with W to describe the magnitude of resources and potential supply, and (2) depth (h) is a component of W as well as a determinant of cost, there is a good reason for its inclusion within the model.

Perhaps the single most important effect of depth is in exploration, for even shallow cover decreases considerably the probability that an isolated deposit will be detected. Consider, for example, the following that describes an index of discoverability (I) of tabular sandstone uranium deposits:

$$I = 4.35 \, q^{0.48} \left(\frac{t}{10^6} \right)^{0.80} h^{-0.64},$$

where   q = deposit average grade in percent
          t = deposit tonnage (s.t.)
          h = depth to deposit in feet

This equation was obtained by statistical analysis of responses of consultants who are experienced and expert in the exploration for sandstone uranium deposits of the western U.S. (Harris and others, 1981; Harris and Chavez, 1984).

Depth clearly increases mining cost, but this effect differs
inversely with deposit size, exerting a smaller, but nevertheless
important, increase in cost with increasing depth for a deposit of
10×10^6 s.t. than for one of 1.0×10^3 s.t. of ore. Table 1 shows
estimated mining cost per ton of ore in \$1974 for tabular
sandstone deposits in the western U.S.

Intradeposit Grade Variance and Transformation of Mineral Deposit to Ore

Mineral resource or exploration models typically describe mineral
occurrences by either a bivariate lognormal distribution for
deposit size and grade, or by two independent lognormal
distributions, one for deposit size and one for deposit grade. Such
a treatment does indeed represent variation in quantity and
quality across the population of deposits. When analysis focuses
on the interaction of economics with endowment to describe
potential supply, modeling that does not include intradeposit
grade variation is too restrictive, for the entire deposit is
considered to be of one grade, that grade generated from the
model of deposit grade. Being that grade generally is not constant
within a deposit, there is a different average grade for each cutoff
grade that is feasible, given mining design, cost relations, and
product price.

Typically, a mining operation does not mine all of the material in
the deposit, because the present value of profits is maximized by
leaving some of the low-grade material in place or on the dump.

As a generalization, profit maximization leads to reduction of
quantity for an increase in quality. Critical to this maximization is
the selection of the appropriate cutoff grade and its associated ore
tonnage and grade.

Thus, it is appropriate to view economics as exerting a
transformation of deposit tonnage and grade to ore tonnage and
grade. Even a simplistic modeling of this transformation requires
a description of the deposit tonnage by grade, and this

Table 1. Estimated mining cost* per ton of ore (as of 1.1.74).

| | Estimated Mining Cost | | | | |
| | - - - - - - - - - Thousands of Tons of Ore - - - - - - - - - | | | | |
Depth (Feet)	0-1	10	100	1,000	$\geq$ 10,000
Open Pit					
0-100	$ 21.00	$ 19.00	$17.00	$15.50	$12.50
150	21.00	19.00	17.00	16.50	12.50
200	22.50	21.50	19.00	16.50	15.50
300	29.00	27.00	21.50	19.50	19.00
Underground					
300	$ 29.00	$ 27.00	$21.50	$19.50	$19.00
400	260.00	45.00	23.50	21.50	21.50
500	400.00	51.00	23.50	21.50	21.50
1,000	1,215.00	133.00	26.50	23.50	21.50
1,500	1,815.00	195.00	31.00	24.50	24.50
2,000	2,615.00	283.00	38.50	24.50	24.50
3,000	4,150.00	405.00	55.00	26.50	26.50
4,000	5,615.00	575.00	70.00	29.00	29.00
$\geq$5,000	7,515.00	780.00	90.00	32.00	32.00

* 1/1/74 estimates include only mine surface plant, equipment, primary development, direct, and indirect mining costs.
(Source: Ellis, Harris, and Van Wie, 1975)

description requires <u>not just deposit tonnage and average grade</u> <u>but the intradeposit grade variance</u>, v.

Figure 3 shows schematically the effect of profit maximization on reported deposit tonnage and grade. Statistical data on known deposit size and grade constitute not only a sample from a population that is biased by exploration and truncated by a cost function, but also one in which the deposits are <u>translated</u>.

Summary of Needed Extensions

The importance of depth (h) as a cost determinant and intradeposit grade variance (v) in mining economics strongly indicates that a four-dimensional (q,t,h,v) pdf is preferred to one that is bivariate (t,q). Of course, if the concerns registered in the previous section about correlation are accommodated, this expanded model must be generalized to a <u>dependent four-</u>

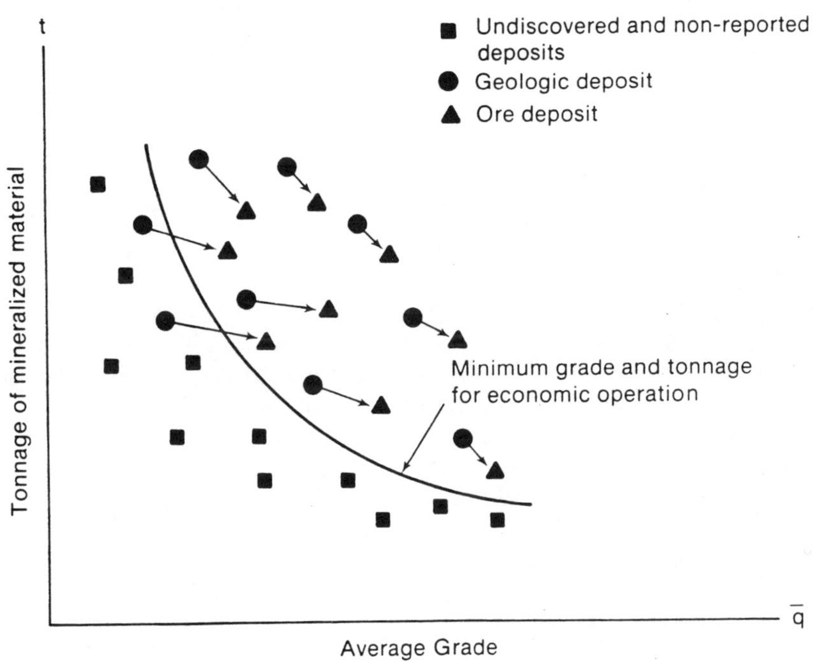

Figure 3. Schematic illustration of truncation and translation resulting from economics of exploration and exploitation.

<u>dimensional</u> model. Finally, accounting for the translation effect requires two transformation functions, one for ore grade and one for ore tonnage, with communication between them.

The following section describes an expanded, comprehensive, crustal abundance model, and that section is followed by one which describes available data and how they were employed to generate model requirements. The paper concludes with the description of preliminary results -- an initial estimate of the expanded model, given rudimentary cost relations and readily available, but limited, deposit data.

A COMPREHENSIVE CRUSTAL ABUNDANCE MODEL FOR U.S. URANIUM

Theory

Consider mineral occurrence to be four dimensional: t, quantity of mineralized rock; q, average grade (%) of t; v, the variance of grade within t (intradeposit grade variance); and h, depth (ft) within the Earth's crust to the occurrence. Then, the foundation for a comprehensive crustal abundance model is the joint pdf, w( ), for t, q, h and v:

$$w\,(\ln t,\, \ln q,\, \ln v,\, h)$$

Following Drew (1977), the preferential discovery by exploration as performed of the largest and richest (highest q) deposits is represented by purposefully biasing w( ) to form a pdf for discoverable deposits - see Charles River Associates, Inc. (1978).[1] However, in this situation, rather than discovery being proportional to t • q, the proportionality is to $q^{0.48}$ • course, unlike Drew's model, this one contains depth to deposit, h, and h was shown to be a strong determinant of discoverability. This depth effect is represented by a weighting function which weights probabilities for occurrence inversely to depth: $0.00175e^{-0.00175h}$. The parameter (0.00175) was derived from the exponent of h in the discoverability function: $q^{0.48}\, t^{0.8}\, h^{-0.64}$ = Index of discoverability.

---

[1]The author originally developed some of these ideas while a consultant to CRA on a research contract for the National Science Foundation.

The pdf for discoverable deposits, w^* ( ), is a biased form of w( ), as indicated by the following proportionality relationship.

$$w^* (\ln t, \ln q, \ln v, h) \quad \propto$$

$$(q^{0.48} t^{0.8}) \bullet w(\ln t, \ln q, \ln v, h) \bullet (0.00175e^{-0.00175h})$$

or, because h is taken to be independent of lnt and lnq, we can write the following:

$$w^* (\ln t, \ln q, \ln v, h) \propto$$
$$\left(q^{0.48} t^{0.8}\right) \bullet w (\ln t, \ln q, \ln v, h) \bullet$$
$$\left(0.00175e^{-0.00175h}\right)$$

Specifically,

$$w^* (\ln t, \ln q, \ln v, h) = f^* (\ln t, \ln q) \bullet y_t (\ln v) \bullet \alpha^* (h) ,$$

where

$$f^* (\ln t, \ln q) = \left[\frac{1}{(2\pi) \sigma_t \bullet \sigma_q \sqrt{1 - \rho^2}} \right] \bullet \exp \left\{ \left[- \frac{1}{2 (1 - \rho^2)} \right] \right.$$

$$\bullet \left[\frac{[\ln t - (\mu_t + \beta_1 \sigma_t^2)]^2}{\sigma_t^2} \right.$$

$$- \frac{2\rho[\ln t - (\mu_t + \beta_1 \sigma_t^2)] \bullet [\ln q - (\mu_q + \beta_2 \sigma_q^2]}{\sigma_t \bullet \sigma_q}$$

$$\left. \left. + \frac{[\ln q - (\mu_q + \beta_2 \sigma_q^2)]^2}{\sigma_q^2} \right] \right\}$$

$$y_t(\ln v) = \left[\frac{1}{\sigma_v\sqrt{2\pi(1-\rho^2)}}\right] \cdot \exp\left\{\left[-\frac{1}{2\sigma_v^2(1-\rho^2)}\right]\right.$$

$$\left.\cdot\left[\ln v - \mu_v - \rho\cdot\sigma_v\left(\frac{\ln t - \mu_t}{\sigma_t}\right)\right]^2\right\}$$

$$\alpha(h) \propto (h)\cdot(0.00175e^{-0.00175h})$$

where $\alpha(h) - 1/DEPTH$, $0 < h \le DEPTH$,
DEPTH is maximum depth to deposit.

Specifically,

$$\alpha^*(h) = \alpha(h)\cdot(0.00175e^{-0.00175h}) \Big/ \int_0^{DEPTH}\alpha(h)\cdot(0.00175e^{-0.00175h})\,dh$$

Thus,

$$\alpha^*(h) = (0.00175e^{-0.00175h}) / (1.0 - e^{-0.00175\,DEPTH}),\quad 0 < h \le DEPTH$$

$$= 0,\text{ otherwise}$$

The pdf $f^*(\ )$ is related to $f(\ )$ as follows:

$$f^*(\ln t, \ln q) \propto q^\beta\, t^\beta\, f(\ln t, \ln q)$$

Because $f(\ )$ is a dependent bivariate normal pdf, $f^*(\ )$ is $f(\ )$ with a shift in its location parameters, that is μ_t and μ_q. Specifically, the parameters of $f^*(\ )$ and $f(\ )$ are related as follows:

$$\mu_q^* = \mu_q + \beta_1\,\sigma_q^2$$

$$\mu_t^* = \mu_t + \beta_2\,\sigma_t^2$$

$$\sigma_q^* = \sigma_q$$

$$\sigma_t^* = \sigma_t$$

As noted previously, $\beta_1 = 0.48$ and $\beta_2 = 0.80$.

Define two transformation functions that describe ore tonnage $(\tilde{t})$ and grade $(\tilde{q})$ as functions of t, q, v, h, and price (p):

$$\ln \tilde{q} = \phi_q \, (\ln t, \ln q, \ln v, h, p)$$

$$\ln \tilde{t} = \gamma \, (\ln t, \ln q, \ln v, \ln \tilde{q}, h, p)$$

By substitution, ln q can be purged from γ ( ), defining a new function ϕ_t ( ) :

$$\ln \tilde{t} = \phi_t \, (\ln t, \ln q, \ln v, h, p)$$

Given ϕ_t ( ) and ϕ_q ( ), ln t and ln q can be described as inverse functions ϕ_t^{-1} ( ) and ϕ_q^{-1} ( ):

$$\ln t = \phi_t^{-1} \, (\ln \tilde{t}, \ln q, \ln v, h, p)$$

$$\ln q = \phi_q^{-1} \, (\ln t, \ln \tilde{q}, \ln v, h, p)$$

Employing the inverse functions of ϕ ( ) and transformation of variables, Ψ (ln $\tilde{t}$, ln $\tilde{q}$, h; p), the joint pdf for ore tonnage, ore grade, and ore depth, given p, is derived from the basic model for discoverable deposits:

$$\Psi(\ln \tilde{t}, \ln \tilde{q}, h; p) =$$

$$\iiint_{QTV} w^*[\phi_t^{-1}(\ln \tilde{t}, \ln q, \ln v, h, p), \phi_q^{-1}(\ln t, \ln \tilde{q}, \ln v, h, p),$$

$$\ln v, h] \ |J| \ dv \ d\ln t \ d\ln q$$

Now, define function $\lambda (\ln t, \ln q, h)$ which describes total unit cost per lb $U_3 O_8$ (c) as a function of ore tonnage and grade and depth to deposit:

$$c = \lambda (\ln \tilde{t}, \ln \tilde{q}, h)$$

Then, combining Ψ ( ) and the inverse functions of λ ( ), given c = p, four model expectations can be described.

$$E[\tilde{Q}|p] =$$

$$\int_{Q} \int_{\lambda_t^{-1}(p, \ln \tilde{q}, h)}^{\infty} \int_{0}^{\lambda_h^{-1}(\ln \tilde{q}, \ln \tilde{t}, p)} \tilde{q} \bullet \Psi(\ln \tilde{t}, \ln \tilde{q}, h; p) \ dh \ d\ln \tilde{t} \ d\ln \tilde{q} \ / K$$

$$E[\tilde{T}|p] =$$

$$\int_{T} \int_{\lambda_q^{-1}(\ln \tilde{t}, p, h)}^{\infty} \int_{0}^{\lambda_h^{-1}(\ln \tilde{q}, \ln \tilde{t}, p)} \tilde{t} \bullet \Psi(\ln \tilde{t}, \ln \tilde{q}, h; p) \ dh \ d\ln \tilde{q} \ d\ln \tilde{t} \ / K$$

$$E[(\tilde{Q} - E[\tilde{Q}|p])^2|p] =$$

$$\int_{Q} \int_{\lambda_t^{-1}(p, \ln \tilde{q}, h)}^{\infty} \int_{0}^{\lambda_h^{-1}(\ln \tilde{q}, \ln \tilde{t}, p)} \tilde{q}^2 \bullet \Psi(\ln \tilde{t}, \ln \tilde{q}, h; p) \ dh \ d\ln \tilde{t} \ d\ln \tilde{q} \ / K$$

$$- \{E[\tilde{Q}|p]\}^2$$

$$E\left[(\tilde{t} - E[\tilde{t} \mid p])) \mid p \right] =$$

$$\int_{\tilde{T}}^{\infty} \int_{\lambda_q^{-1}(\ln \tilde{t},p,h)}^{\lambda_h^{-1}(\ln \tilde{q},\ln \tilde{t},p)} \int_{0} \tilde{t}^{2} \cdot \Psi(\ln \tilde{t}, \ln \tilde{q}, h; p) \, dh \, d\ln \tilde{q} \, d\ln \tilde{t}/K$$

$$- \left\{ E\left[\tilde{t} \mid p \right] \right\}^{2}$$

where

$$K = \int_{\tilde{Q}}^{\infty} \int_{\lambda_t^{-1}(p,\ln \tilde{q},h)}^{\lambda_h^{-1}(\ln \tilde{q},\ln \tilde{t},p)} \int_{0} \Psi(\ln \tilde{t}, \ln \tilde{q}, h; p) \, dh \, d\ln \tilde{t} \, d\ln \tilde{q}$$

Suppose that from actual data on ore deposits, statistics (q, t, s_q^2, s_t^2) can be computed that are comparable in concept to the expectations. Then, define an error measure, Δ:

$$\Delta = \left(1 - \frac{E[\tilde{Q} \mid p]}{\tilde{q}} \right)^{2} + \left(1 - \frac{E[\tilde{T} \mid p]}{\tilde{t}} \right)^{2} + \left(1 - \frac{E[(\tilde{Q} - E[\tilde{Q} \mid p])^{2} \mid p]}{s_{\tilde{q}}^{2}} \right)^{2}$$

$$+ \left(1 - \frac{E[(\tilde{T} - E[\tilde{T} \mid p])^{2} \mid p]}{s_{\tilde{t}}^{2}} \right)^{2}$$

Of course, each of these expectations changes if the value of any one of the four[2] parameters changes; consequently, it is useful

---

[2]Although f(ln t, ln q), a dependent bivariate pdf, requires five parameters, only four need estimation, for two of the five are related through crustal abundance (A): $\ln A = \mu_q + \sigma_q^2/2$.

conceptually to consider the measure of error and the expectations to be functions of the parameter

$$\Delta = E(\sigma_q, \mu_t, \rho) =$$

$$\left[1 - \frac{\varepsilon_q(\sigma_q, \mu_t, \sigma_t, \rho)}{\tilde{\tilde{q}}} \right]^2 + \left[1 - \frac{\varepsilon_t(\sigma_q, \mu_t, \sigma_t, \rho)}{\tilde{\tilde{t}}} \right]^2$$

$$+ \left[1 - \frac{\varepsilon_{s_q}(\sigma_q, \mu_t, \sigma_t, \rho)}{s_{\tilde{q}}^2} \right]^2 + \left[1 - \frac{\varepsilon_{s_t}(\sigma_q, \mu_t, \sigma_t, \rho)}{s_{\tilde{t}}^2} \right]^2$$

With this functional description as a reference, the approach of this study is to use a generalized computer search routine to determine the values of σ_q, μ_t, σ_t, and ρ that minimize function $E(\ )$.

Overview of Requirements of the Model

A version of the crustal abundance model was programmed to support the estimation of the unknown parameters σ_q, μ_t, σ_t, and ρ. Required inputs include the following:

A

y_t (ln v)

$\alpha(h)$

$q^{\beta_1} \cdot t^{\beta_2}$

ke^{-bh}

ϕ_t (ln t, ln q, ln v, h, p)

ϕ_q(ln t, ln q, ln v, h, p)

λ (ln t̃, ln q̃, h)

p

$$\tilde{\bar{t}}$$
$$\tilde{\bar{q}}$$
$$s_t^2$$
$$s_{\tilde{q}}^2$$

Of these, only the following were readily available from previous studies or routinely computable:

$$A = 0.0003 \% \ U_3O_8$$
$$\beta_1 = 0.46$$
$$\beta_2 = 0.80$$
$$k = 0.0019065$$
$$b = 0.00175$$
$$\bar{t} = 656{,}908 \ s.t.$$
$$s_t^2 = 3503 \times 10^9$$
$$\bar{q} = 0.184\% \ U_3O_8$$
$$s_{\tilde{q}}^2 = 0.0161$$
$$p = \$50.00/lb \ U_3O_8$$

β_1, β_2, k, and b were obtained from a previous study on uranium exploration, which produced a discoverability relation (Harris and Chavez, 1984). Statistics $\tilde{\bar{t}}$, s_t^2, $\tilde{\bar{q}}$, and $s_{\tilde{q}}^2$ must be calculated from data.

Data and Relations for this Demonstration

Deposit tonnage and grade data for this demonstration were taken from the $50 preproduction (reserves plus cumulative production) inventory published by U.S. DOE (1978). Figure 4 and Table 2 present the basic DOE data used for the calculation of statistics. Clearly, the data of Table 2 are less than ideal because of the few large class intervals used to depict variation in deposit size. Nevertheless, they will be used for this demonstration.

Lacking data or model for $\alpha(h)$, it is <u>assumed</u> to be rectangular over the interval of 0 to 2000 ft; therefore,

$$\alpha(h) = 1/2000, \ 0 \le h \le 2000$$
$$= 0, \ otherwise$$

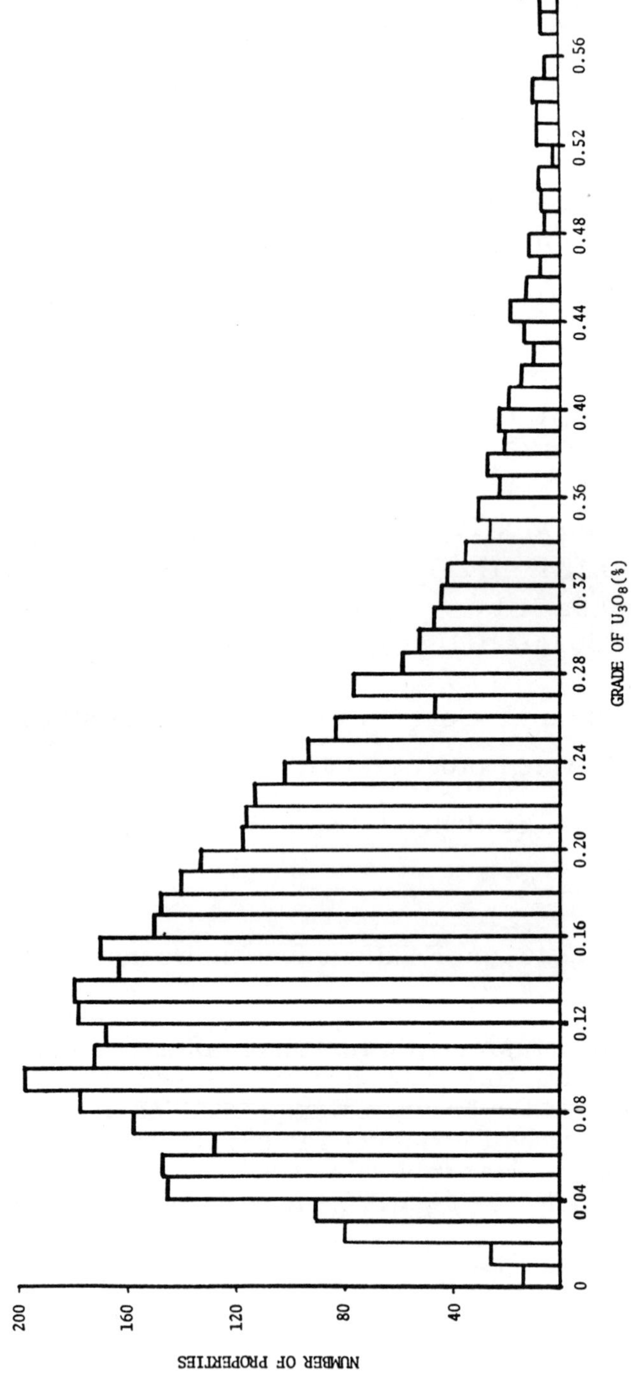

Figure 4. U.S. distribution of $50 reserves plus production by grade increment as of 1/1/78.

Table 2. Frequency of deposits in $50 preproduction inventory by tonnage and grade classes.

Ore (x 1000 s·t)	Average Grade ($\%U_3O_8$)				
	0 – 0.05	0.05 – 0.10	0.10 – 0.20	0.20 – 0.40	>0.40
>8000	12	26	10	0	0
4000 – 8000	15	20	9	0	0
2000 – 4000	18	21	10	2	0
1000 – 2000	21	32	16	4	0
500 – 1000	16	29	10	6	0
0 – 500	73	325	714	382	24

Source: U.S. Department of Energy (1978)

The remaining inputs were not available; consequently, O(accommodating the model's input requirements required considerable supplementary analysis to obtain usable estimates. Particularly problematic are the ore transformation functions and a probability distribution for intradeposit grade variance. Although all of these clearly are desirable in resource and potential supply models, data for their estimation either are nonexistent or unavailable. Indirect approaches were employed to estimate or model these components.

The probability distribution for v, conditional upon t, was estimated by Monte Carlo analysis, using (1) the DOE (1978) distribution of number of deposits by grade (see Fig. 4) and the frequency of deposit size for grade classes (see Table 2), (2) subjective information on sample size for grade determination, and (3) the Matheron-DeWijs variance-volume relationship. This analysis produced the histogram for ln v of Figure 5. Because a correlation was identified between ln v and ln t, this histogram represents the marginal distribution for ln v. As a result of this analysis, the following model was defined and estimated for y_t(ln v), the conditional pdf for v:

$$y_t(\ln v) = \left[\frac{1}{\sigma_v \sqrt{2\pi(1 - \rho^2)}} \right] \cdot \exp \left\{ \left[- \frac{1}{2\sigma_v^2 (1 - \rho^2)} \right] \right.$$

$$\left. \cdot \left[\ln v - \mu_v - \rho\sigma_v \left(\frac{\ln t - \mu_t}{\sigma_t} \right) \right]^2 \right\},$$

where $\hat{\mu}_v = -1.65$, $\hat{\sigma}_v = 0.685$, and $\hat{\rho} = 0.65$.

Substituting these values into the given equation, we can write y_t(ln v) as follows:

$$y_t(\ln v) = \left(\frac{1}{2.486} \right)$$

$$\cdot \exp \left\{ (-1.845) \left[\ln v - (-1.65) - 0.445 \left(\frac{\ln t - \mu_t}{\sigma_t} \right) \right]^2 \right\},$$

$$-\infty \leq \ln v \leq \infty$$
$$-\infty \leq \ln t \leq \infty$$

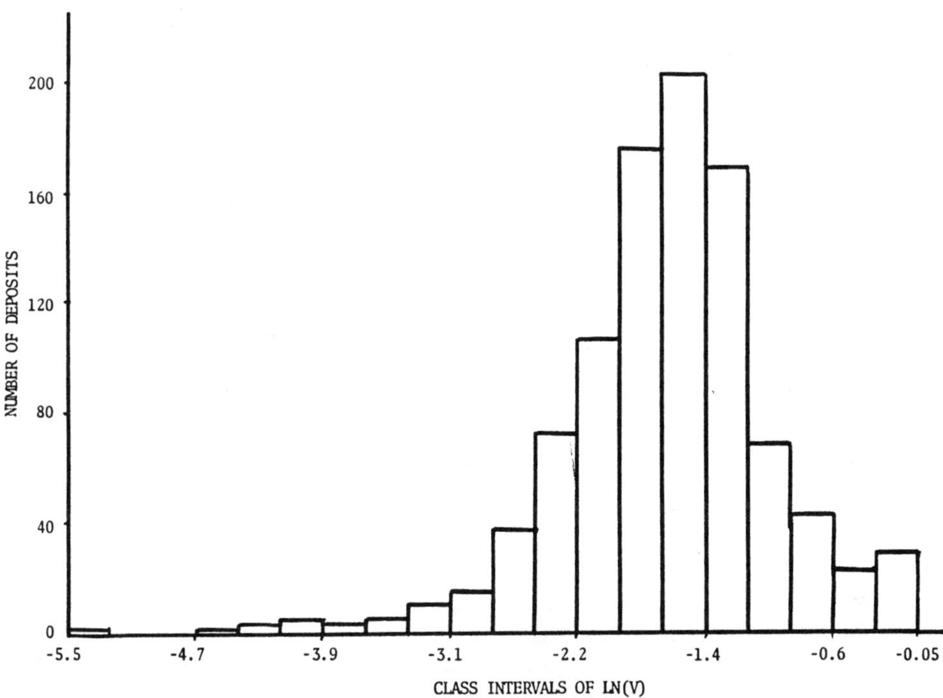

Figure 5.   Histogram of ln(v) produced by Monte Carlo analysis.

The nonavailability of ϕ_t ( ) and ϕ_q ( ), ore transformation
functions, and of data for their estimation forced the design of a
different approach to representing translation and truncation due
to economics of profit maximization. Specifically, a computer
subroutine for mine optimization was designed as an integral part
of the computerized crustal abundance system. This subroutine
takes as input the deposit characteristics (t, q, h, and v) and
determines that cutoff grade for which the net present value of
profits is a maximum. The ore tonnage (t) and grade (q)
associated with maximum net present value are included in the
model's inventory, given that maximum net present value is
positive. Those deposits having a negative maximum net present
value are considered to belong to the truncated part of the deposit
population, that is that part which either is not known or not
reported by industry.

The foundation for determining net present value consists of
price (p), which is an input value, and a relation that describes c,
production cost per lb $U_3 O_8$. In actual practice, production cost

reflects a large number of physical and chemical features that interact to influence mining and processing costs.

Furthermore, the ore tonnage and average ore grade reported for an actual mine result from (1) examining the ways that these features interact to affect production cost, and (2) selecting that production design, capital investment, scale of operation, life of operation, and cutoff grade that maximizes the net present value of the deposit to the resource holder.

In this demonstration, no attempt is made to model comprehensively the many features of a deposit or the complexity of the optimization decision. Instead, a deposit is described by only t, q, v, h and their correlations; costs are related only to these features; and mine optimization is achieved by searching for the cutoff grade that maximizes net present value, given these simple relations.

A final simplification is that the relationship of mining cost to depth and deposit size for tabular sandstone deposits of the San Juan Basin was used for all deposits. Clearly, such a simplification ignores significant differences in morphologies, grade distribution, and cost relations that exist between tabular sandstone deposits and other deposit types, for example the shallow, roll-type deposits of Wyoming. This simplistic approach undoubtedly introduces distortions; the only defense for using it is that it makes possible an early demonstration of the approach. Naturally, resulting parameter estimates must be viewed with caution, as initial rough estimates, estimates to be replaced when better data and relations are available.

Specifically, in this demonstration production cost (c) in dollars/lb U_3O_8 was described by the following equation:

$$c = \frac{E \bullet m(\tilde{t}, h) + 7.181 \bullet E}{19.2\ \tilde{q}} + 1.6045E \ ,$$

where E is a cost escalation factor which adjusts costs in \$1974 to costs denominated in dollars of the year of interest (for 1978 E = 1.36), and m(t, h) is functional notation for cost estimates made by DOE in1974 and reported in Table 1 for specific combinations of tonnage and depth.

Production cost, c, consists of operating (op) and capital (cap) costs:

$$c = cap + op$$

Because cap and op are treated differently in discounted cash flow accounting and evaluation, they were separated for the computation of net present value. Here this separation is made simply by defining cap as a constant fraction (α) of c:

$$cap = \alpha c$$

Discounted cash flow analysis also requires mine life, n, and annual production rate of ore, x. Both of these are defined as functions of ore tonnage, t:

$$n = 0.1385 \; t^{0.27}$$

Then, x, annual production, is t/n:

$$x = 7.22 \; t^{0.73}$$

Suppose for expository purposes that a deposit within the crustal abundance system were described by t, q, h, and v. Then, for mining cutoff grade g', ore tonnage $\tilde{t}$ and grade $\tilde{q}$ are computed as follows:

$$\mu_g = \ln q - v/2$$

$$\tilde{t} = t \cdot \left[1 - F\left(\frac{\ln g' - \mu_g}{\sqrt{v}} \right) \right]$$

$$\tilde{q} = e^{\mu_g + v/2} \left[\frac{1 - F\left(\dfrac{\ln g' - (\mu_g + v)}{\sqrt{v}} \right)}{1 - F\left(\dfrac{\ln g' - \mu_g}{\sqrt{v}} \right)} \right]$$

and, given $\tilde{t}$ and $\tilde{q}$, production cost, c, and net present value, NPV, can be computed:

$$c = ((E \bullet m(\tilde{t}, h) + 7.181E) / 19.2\tilde{q}) + 1.6045E$$

$$NPV = \{(7.22\tilde{t}^{\,0.73} \bullet \tilde{q} \bullet R \bullet (20)) \bullet [p - (1 - \alpha) \bullet c] - TX\} \bullet$$

$$\bullet \left\{ \frac{(1 + v)^{0.1385\tilde{t}^{0.27}} - 1}{(1 + v)^{0.1385\tilde{t}^{0.27}} \quad (v)} \right\} - \{t \bullet \tilde{q} \bullet \alpha \bullet c \bullet (20)\}$$

where
- p = price in \$/ lb $U_3 O_8$
- R = overall recovery, as a fraction
- v = required rate of return as a fraction
- TX = income tax, which is computed according to accepted income tax accounting, allowing for depletion and depreciation.

The equation shows NPV to be a function of $\tilde{t}$, $\tilde{q}$, and h, given R, p, v, t, q, h, v, and tax accounting parameters. More specifically, given these values, NPV is a function of cutoff grade g':

$$NPV = \beta(g' ; R, p, v, t, q, h, v, ...)$$

This relation presents profit maximization as the selection of that value for g' that maximizes function $\beta(\ )$. Accordingly, within the crustal abundance system, cutoff grade (g') was changed in a systematic fashion, computing NPV for each grade and identifying that cutoff grade for which NPV was a maximum. Values of some of the parameters or conditions for the cash-flow analysis that supports the calculation of NPV are the following:

- p = \$50.00 / lb $U_3 O_8$
- v = 0.15
- % depletion rate = 15%
- maximum depletion charge $= 0.5 \bullet (p - c) \bullet (7.22t^{0.72})$
  $\bullet \tilde{q} \bullet R \bullet (20)$
- straight-line depreciation

R = 0.85
E = 1.36
tax rate = 0.5

Estimates of Parameters

The crustal abundance system was linked to a general search
program which searched the four-dimensional parameter space
for those values which, when employed by the system, produced
expectations that approximated the statistics for actual ore
deposits. This search yielded the following parameters for
f(ln t,ln q):

$$\mu_q = -9.1491$$

$$\sigma_q{}^2 = 2.0747$$

$$\mu_t = 11.7065$$

$$\sigma_t{}^2 = 0.60477$$

$$\rho = -0.25$$

$$\text{Given } A = 0.0003 = e^{\mu_q + \sigma_q^2/2}$$

Associated with these parameters are the expectations of the
model for ore deposit characteristics, which compare well with
the actual statistics from ore deposit data, as shown in Table 3.

It is noteworthy that the parameters of the grade distribution
produced by this model and analysis are nearly identical to those
estimated by Harris and Chavez (1981) using the ratio of stocks
$(w_{q'}/W)$ and employing an asymptotic variance estimate:
ln Q ~ N(-9.1442, 2.065). Of course, additional insight about the
sizes of mineral occurrences and the relationship of size to grade
has been gained and incorporated in the model. The impact of
this gain is appreciated fully only when the crustal abundance

Table 3. Comparison of model expectations with deposit
          statistics.

	Statistics of Known Deposits	Expectations of the Model
$\bar{q}$	0.184%	0.192%
s_q	0.1269	0.1256
$\bar{t}$	656,908	715,388
s_t	1.8716×10^6	1.153×10^6

models are used to support economic analysis and estimation of
mineral resources and potential supply. Such analysis now is
feasible, whereas a simple model that does not include deposit
size and depth could not be used similarly without making strong
assumptions or without auxiliary relations.

Summary Description of the Comprehensive Model

The comprehensive model of uranium occurrence in the United
States that emerges from this study is the following:

$$w (\ln t, \ln q, \ln v, h) = f (\ln t, \ln q) \cdot y_t (\ln v) \cdot \alpha (h)$$

where $f (\ln t, \ln q) \sim N (\mu_q, \sigma_q^2, \mu_t, \sigma_t^2, \rho)$,

$$\text{where} \quad \mu_q = -9.1491$$
$$\sigma_q^2 = 2.0747$$
$$\mu_t = 11.7065$$
$$\sigma_t^2 = 0.60477$$
$$\rho = -0.25$$
$$A = 0.0003 = e^{\mu_q} + \sigma_q^2/2 \ = \text{mean crustal concentration in \%}$$

$$y_t (\ln v) \sim N (\mu_{v/t}, \sigma_{v/t}^2) ,$$

where $\mu_{v/t} = -1.65 + 0.445((\ln t - 11.7065)/0.778)$,

$$0.445 = (\rho_{t,v})(\sigma_v) = (0.65)(0.685),$$
$$\text{and}$$
$$\sigma_{v/t}^2 = \sigma_v^2(1 - \rho_{t,v}^2) = (0.469)(0.5775) = 0.27$$

$\alpha(h) \sim R(0, 2000)$

$\alpha(h) = 1/2000, \quad 0 \leq h \leq 2000$

$\alpha^*(h) = 0.0019065e^{-0.00175h}, \quad 0 \leq h \leq 2000$

$f^*(\ln t, \ln q) = N(\mu_q^*, \sigma_q^2, \mu_t^*, \sigma_t^2, \rho)$,

where

$$\mu_q^* = \mu_q + \beta_1 \sigma_q^2 = -9.1491 + (0.8)(2.075) = -7.4891$$

$$\sigma_q^2 = 2.075$$

$$\mu_t^* = \mu_t + \beta_2 \sigma_t^2 = 11.7065 + (0.46)(0.605) = 11.9848$$

$$\sigma_t^2 = 0.605$$

$$\rho = -0.25$$

Finally,

$$w^*(\ln t, \ln q, \ln v, h) = f^*(\ln t, \ln q) \bullet y_t(\ln v) \bullet \alpha^*(h)$$

$$c = \frac{E \bullet m(\tilde{t}, h) + 7.181E}{19.2 \, \tilde{q}} + 1.6045E$$

$$NPV = \left\{ (7.22t^{0.73}q \cdot R(20))[p-(1-\alpha)c] - TX \right\}$$

$$\cdot \left\{ \frac{(1+v)^{0.1385\tilde{t}^{0.27}} - 1}{(1+v)^{0.1385\tilde{t}^{0.27}}(v)} \right\} - \left\{ \tilde{t} \cdot \tilde{q} \cdot \alpha \cdot c(20) \right\}$$

Comparison

A comprehensive comparison of model and data, a natural and in most situations useful exercise, must await the completion of the potential supply system. Only then will it be possible to compare the model's U_3O_8 potential supply to the U.S. inventory of cumulative production plus reserves. This is an especially important comparison in this situation, because parameters of the model were estimated statistically, ignoring the relationships of stock to statistics. On a more limited, but, nevertheless important basis, the model has been shown to be conformable with ore deposit statistics (Table 3).

Another important but limited comparison is the model's U_3O_8 endowment and the DOE inventory of cumulative production plus reserves. Because DOE's inventory basically is a description by grade of potential supply for a forward cost of \$50/lb U_3O_8 it is not comparable strictly to U_3O_8 endowment. Figure 6 shows these two stocks and their divergence as cutoff grade is lowered. Of course, the proposition of this study is that when size, grade, depth, cost, and price are allowed for, this endowment is compatible with DOE's stock. It is satisfying that the contradiction of an earlier model, based upon stock analysis only, of model endowment being less than the known stock for the higher grades, is not present in this model.

Table 4 shows the parameters of the various CA models that have been developed for U.S. uranium. As duly noted earlier, the structures of these models differ, diminishing the usefulness of simple comparisons of parameters.

Perhaps, a more interesting exercise is to compare estimates of endowment by the different models. To that end, suppose that each bivariate model is represented by its marginal distribution of

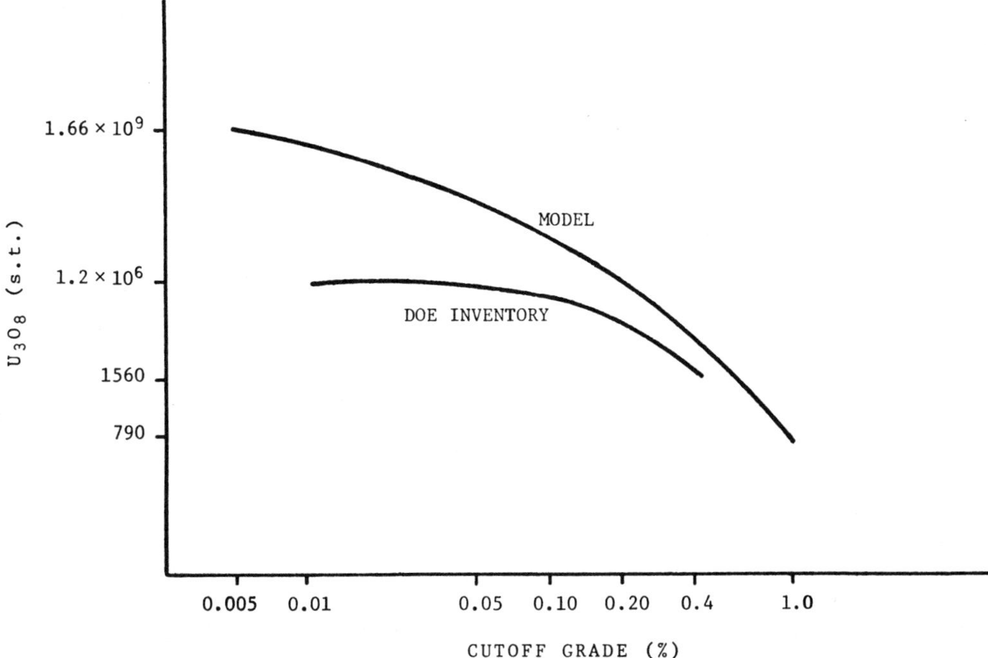

Figure 6. U_3O_8 endowment (model) and U.S. DOE inventory.

grade, that is it describes U_3O_8 endowment in terms of
concentration only. Employing this procedure, endowment was
estimated by each of four crustal abundance models. Table 5
shows these estimates as quantity and average grade of
endowment for cutoff grades of 0.10 and 0.20 % U_3O_8.

The estimates in Table 5 take on more meaning when they are
considered in light of the chronology of the methodologies by
which they were estimated. In this regard, note that the quantity
of endowment (0.37 x 10^6 s.t.) estimated by Harris and Chavez
using an average of estimated σ for high grades is less than the
DOE inventory, but average grades match closely those of the
inventory. Because of this anomaly, Harris and Chavez next

Table 4. Parameters of crustal abundance models of U.S. uranium
         endowment.

	Harris and Chavez		Drew	This Study
	Stock	Asym. $\hat{\sigma}$		
$\bar{q}$	−9.0312	−9.1442	−10.0917[*]	−9.1491
σ^2_q	1.839	2.065	3.96	2.075
$\bar{t}$	−	−	7.517	11.7065
σ^2_t	−	−	1.362	0.605
ρ	−	−	−	−0.25

* In Drew's analysis grade is a proportion; consequently, the value
reported (-15.10) has been adjusted to % equivalent. Additionally,
Drew used 0.0002, instead of 0.0003, for crustal concentration.
The value of -10.0917 is a value adjusted for both effects.

employed an asymptotic value of estimated σ; quantities of
endowment by this model are greater than the inventory
quantities for both cutoff grades, which is a desired result, but
now the average grades do not match nearly so well those of the
inventory.

Drew's bivariate model, which was the first attempt to model both
concentration and deposit size and to estimate parameters by a
statistical approach, produces endowments that are large when
compared to the DOE inventory and the other model estimates,
but average grades compare reasonably well with those of the
inventory.

Table 5. Comparison of model estimates of U_3O_8 endowment.

| Cutoff Grades (%) | DREW | | HARRIS AND CHAVEZ | | | | THIS STUDY | |
| | | | $\bar{\sigma}$ | | Asym. $\hat{\sigma}$ | | | |
	U_3O_8 ($\times 10^6$)	Ave. Grade	U_3O_8 ($\times 10^6$)	Ave. Grade	U_3O_8 ($\times 10^6$)	Ave. Grade	U_3O_8 ($\times 10^6$)	Ave. Grade
0.20	57	0.40	0.37*	0.33	0.9	0.27	1.1	0.33
0.10	152	0.17	2.6	0.18	5.4	0.14	2.9	0.19

U.S. DOE INVENTORY

	U_3O_8 ($\times 10^6$ s.t.)	Ave. Grade (% U_3O_8)
0.20	0.43*	0.36
0.10	0.74	0.21

The univariate form of the model of this study produces quantities
of endowment more similar to those of the asymptotic-$\hat{\sigma}$ model
than those of Drew's model. Average grades of endowment
compare more closely to those of the inventory than do estimates
of the other three models. The close match of expectations of the
full model with statistics of ore properties, the close agreement of
endowment average grades for specified cutoff grades, and the
fact that quantity of endowment exceeds considerably the known
inventory for all grades gives this crustal abundance model
considerable appeal.

Of course, any comparison with Drew's model must allow for the
following differences in models:

> Drew used \$8.00/lb $U_3 O_8$ and statistical data for January 1,
> 1970.

> Crustal abundance in Drew's model is a proportion instead
> of a percentage.

> Drew's model does not consider depth.

In Drew's model deposit tonnage and grade are
independent statistically.

Drew's model does not allow for the fact that statistical data
are for the ore deposit not the mineral deposit, and no
provision is made for translation.

Intuitively, predicating the crustal abundance model of this study
upon concentration as a proportion would have decreased the
differences in $\hat{\mu}_q$. Because such an adjustment would be
equivalent to adding the natural logarithm of 0.01. to all
concentrations, the estimate of $\hat{\sigma}_q^2$ would be unchanged because
this same constant also would be added to the mean, leaving the
second moment unchanged; however, the estimate of μ_q would be
changed to -13.754, a value closer to the estimate by Drew. Of
course, significant differences remain, which is not surprising
when differences in model structure and complexity and in data
are considered. Finally, comparison of crustal abundance models
solely by examination of parameters and endowment estimates
does not recognize the estimation of parameters through systems
that include economic activities and relations. Consequently,
comparisons of models should include an examination of the
resources and potential supply that are described by each system
for comparable economic circumstances. Such comparisons have
not yet been made, but efforts to do so have been initiated.

SUMMARY STATEMENT

The results reported from this study should be considered
preliminary and subject to revision as better data are assembled.
These results undoubtedly reflect compromises made because of
the state of information available at the time of the initial
investigation. For example, the representation of mining costs by
depth and size was based upon a simplified cost analysis reported
in 1974 for deposits of the San Juan Basin of New Mexico.
Clearly, there are some differences in costs among regions and
deposit types, for example the roll-type sandstone deposits of
Wyoming and Texas in contrast to the tabular sandstone deposits
of the San Juan Basin.

Everything else being equal, credibility of a crustal abundance
model is diminished by representing costs of all modes of

occurrences by costs of a single deposit type. Effort at present is being made to model costs more comprehensively.

One of the motivations for crustal abundance modeling is the estimation of resources and potential supply as a function of price and other economic variables. Only when this has been done will the implication of the model be clearly revealed. Such work already is underway, but results are not available yet.

ACKNOWLEDGMENT

I wish to acknowledge the contribution of Stephen Styers, a PhD student in the mineral economics program of the Department of Mining and Geological Engineering of the University of Arizona, for technical assistance and computer programming for the systems estimation of parameters. Similarly, acknowledgment is made of the contribution of Bahun Tetevi Wilson, a recent MS graduate in mineral economics, for computer programming of the Monte Carlo estimation of the probability density function for intradeposit grade variance. Computer funds for this research were provided by the University of Arizona.

REFERENCES

Agterberg, F.P., 1980, Lognormal models for several metals in selected areas in Canada, in Guillemin, C., and Lagny, P., eds., Colloque C-1. Ressources Minérales-Mineral Resources: 26th CGI, Bur. Recherches Géol. et Minières Mém. 106, Orleans, France, p. 83-90.

Agterberg, F. P., and Divi, S. R. 1978, A statistical model for the distribution of copper, lead, and zinc in the Canadian Appalachian Region: Econ. Geology, v. 73, no. 2, p. 230-245.

Ahrens, L. H., 1953, A fundamental law of geochemistry: Nature, v. 172, no. 4390, p. 1148.

Ahrens, L.H., 1954, The lognormal distribution of the elements: Geochim. Cosmochim. Acta, v.5, no. 2, p. 49-73; v. 6, no. 2/3, p. 121-131.

Brinck, J.W., 1967, Note on the distribution and predictability of mineral resources: Euratom 3461e, Brussels, 25 p.

Charles River Associates, 1978, The economics and geology of mineral supply: an integrated framework for long-run policy analysis: CRA Report No. 327, 468 p.

Deffeyes, K.S., and MacGregor, I.D., 1980, World uranium resources: Sci. American, v. 242, no. 1, p. 66-76.

DeNapoli, F.,1976, Mineral resource adequacy: unpubl. masters thesis, The Pennsylvania State Univ., 217 p.

DeWijs, H.J., 1951, Statistics of ore distribution part 1: Geol. Mijnbouw, v. 30, no. 11, p. 365-375.

DeWijs, H.J., 1953, Statistics of ore distribution, part 2: Geol. Mijnbouw, v. 32, no. 1, p. 12-24.

Drew, M.W., 1977, U.S. uranium deposits - a geostatistical model: Resources Policy, v. 3, no. 1, p. 60-71.

Ellis, J.R., Harris, D.P., and Van Wie, N.H., 1975, A subjective probability appraisal of uranium resources in the State of New Mexico: Open File Report GJO-110 (76), U.S. Energy Research and Development Administration, Grand Junction, Colorado, 97 p.

Erickson, R.L., 1973, Crustal abundance of elements, and mineral reserves and resources, in Brobst, D.A., and Pratt, W.P., eds., United States mineral resources: U.S. Geol. Survey Prof. Paper 820, p. 21-25.

Harris, D.P., 1984a, Mineral resources appraisal: Oxford Univ. Press, New York, 448 p.

Harris, D.P., 1984b., Mineral resources appraisal and policy - controversies, issues, and the future: Resources Policy, v. 10, no. 2, p. 81-100.

Harris, D.P., and Agterberg, F.P., 1981, The appraisal of mineral resources: Economic Geology Seventy-Fifth Anniversary Volume 1905-1980, p. 897-938.

Harris, D.P., and Chavez, L., 1981, Crustal abundance and a potential supply system. Part II, Systems and economics for estimation of uranium potential supply: Research report prepared under subcontract 78-238-E,Open File Report, U.

S. Department of Energy, Grand Junction Office, Colorado,
     p. 385-506.

Harris, D.P., and Chavez, L., 1984, Modelling dynamic supply of
     uranium -- an experiment in the integration of economics,
     geology, and engineering: 18th Intern. Symp. on Application
     of Computers and Mathematics in the Mineral Industry,
     Institution of Mining and Metallurgy, London, p. 817-892.

Harris, D.P., Ortiz-Vertiz, S.R., Chavez, M.L., and Agbolosoo, E.K.,
     1981,Systems and economics for the estimation of uranium
     potential supply: Research report prepared under
     subcontract 78-238-E. Open File Report, U.S. Department of
     Energy, Grand Junction Office, Colorado, 609 p.

Harris, D.P., and Skinner, B.J., 1982, The assessment of long-
     term supplies of minerals, in Smith, V.K., and Krutilla, J.V.,
     eds., Explorations in natural resource economics: The Johns
     Hopkins Univ. Press, Baltimore, Maryland, p. 247-326.

Matheron, G., 1971, The theory of regionalized variables and its
     applications: Les Cahiers du Centre de Morphologie
     Mathematique de Fontainebleau, No. 5, Fontainebleau,
     France, 211 p.

McKelvey, V.E., 1960, Relation of reserves of the elements to
     their crustal abundance: Am Jour. Sci., Bradley Volume,
     v. 258-A, p. 234-241.

Singer, D.A., Cox, D.P., and Drew, L.J., 1975; Grade and tonnage
     relationships among copper deposits: U.S. Geol. Survey Prof.
     Paper 907-A, p. A1 – A11.

Singer, D.A., and DeYoung, J.H.,Jr., 1980, What can grade-tonnage
     relations really tell us? in Guillemin C. and Lagny, P.,eds.,
     Colloque C-1, Ressources Minerales - Mineral Resources:
     26th CGI, Bur. Recherches Geol, et Minieres Mem. 106,
     Orleans, France, p. 91-101.

Skinner, B.J., 1976, A second iron age ahead?: Am. Scientist,
     v. 64, no. 3, p. 258-269.

United States Department of Energy, 1978, Statistical data of the
     uranium industry, January 1, 1978: U. S. Department of
     Energy, Grand Junction Office, Colorado, Open File Report
     GJO-100(78), 91 p.

USE OF DECISION THEORY FOR PATTERN RECOGNITION IN GEOLOGY

M. T. Abasov, I. S. Djafarov, and N. M. Djafarova

Azerbaijan Academy of Sciences

ABSTRACT

Decision theory may be used in geology in situations where it is necessary to make a selection between alternatives. The single calculation of the risk function is based on (1) the matrix of a priori probabilities, (2) an experimental matrix, and (3) loss-matrix. Difficulties in the procedure can be minimized by using a method of "group calculation of arguments" (MGCA). A detailed description of the MGCA algorithm is given along with an example of a solution of a problem.

BASIC CONCEPTS USED IN DECISION THEORY

In its broader sense decision theory studies the problem of determining an optimal decision when it is necessary to make a selection between different alternatives (Truxal and Padalino, 1961).

Suppose that there exist N classes R_1, R_2, ..., R_N. Let one observation be characterized by n features $x = x_1$, x_2, ..., x_n, each feature x_i, i = 1, 2, ..., n taking k discrete values. There is a variety of possible selections $d = \{d_1, d_2, ..., d_N\}$; $d_j = d_j(x) = j$ referring the observation $x = x_1, x_2$, ..., x_n to a class with number j.

One should create a decision rule that will refer the current observation X to one of the classes R_i, i = 1, 2, ..., N with minimum risk.

The decision rule is defined by the
       (1)   experimental matrix **E**;
       (2)   a priori probabilities described by the row-matrix **P**; and
       (3)   loss-matrix **L**.

The experimental matrix **E** gives probabilities of distribution according to the classes.

	$\overline{X}_1$	$\overline{X}_2$	...	$\overline{X}_M$
R_1	P_{11}	P_{12}	...	P_{1M}
R_2	P_{21}	P_{22}	...	P_{2M}
$\vdots$	$\vdots$	$\vdots$	...	$\vdots$
R_N	P_{N1}	P_{N2}	...	P_{NM}

where $M = k^n$ (because each of n features can take k values). P_{2M} is a probability of referring the observation x_M to the class R_2. A priori probabilities $P(R_i)$ are those defining a possibility of appearance of each class.

$$P = \{ P(R_1),\ P(R_2),\ ...,\ P(R_n) \}$$

The loss-matrix **L** determines relative value for the possible errors made when referring the current observation to a certain class.

	d_1	d_2	$\cdots$	d_M
R_1	l_{11}	l_{12}	$\cdots$	l_{1M}
R_2	l_{21}	l_{22}	$\cdots$	l_{2M}
$\vdots$	$\vdots$	$\vdots$	$\cdots$	$\vdots$
R_N	l_{N1}	l_{N2}	$\cdots$	l_{NM}

The element l_{2N} in the given matrix defines the values of losses connected with taking the decision d_N on allocating observations from the class R_2 to the class R_N.

Let us determine expected losses.

Average losses for class R_i for different means x_j are:

$$p(R_i) \;=\; L\left[\, R_i\,,\, d(\bar{x}_1)\,\right] \cdot P\left(\frac{\bar{x}_1}{R_i}\right) \;+$$

$$L\left[\, R_i\,,\, d(\bar{x}_2)\,\right] \cdot P\left(\frac{\bar{x}_2}{R_i}\right) \;+$$

$$\bullet \quad \bullet \quad \bullet \quad \bullet \quad \bullet \qquad +$$

$$L\left[\, R_i\,,\, d(\bar{x}_M)\,\right] \cdot P\left(\frac{\bar{x}_M}{R_i}\right) \;,$$

$$p(R_i) \;=\; \sum_{j=1}^{M} L\left[\, R_i\,,\, d(\bar{x}_j)\,\right] \cdot P\left(\frac{\bar{x}_j}{R_i}\right) \;, \qquad (1)$$

$$i = 1,\, 2,\, ...,\, N.$$

Expression (1) determines the expected value of losses for each class R_i (depending on the decision function d which is not yet selected). The expected value of losses p in all the classes is equal to the weighted average of $p(R_i)$:

$$p = \sum_{i=1}^{N} p(R_i)\ P(R_i) \tag{2}$$

where $P(R_i)$ is a priori probabilities of classes R_i. Substituting (1) for (2) we get

$$p = \sum_{i=1}^{N} P(R_i) \sum_{j=1}^{M} L\left[\ R_i\ ,\ d(\bar{x}_j)\ \right] \bullet P\left(\frac{\bar{x}_j}{R_i}\right) ,$$

$$p = \sum_{j=1}^{M}\left[\ \sum_{i=1}^{N} P(R_i)\ L\left[\ R_i\ ,\ d(\bar{x}_j)\ \right] \bullet P\left(\frac{\bar{x}_j}{R_i}\right)\right] \tag{3}$$

Let us consider the expression in square brackets in Equation (3). Here, $d(x_j)$ is a solution allocating the current observation x_j to one of N classes. Let this be a class j. Then the risk corresponding to this solution, $p(d_j)$, is calculated according to the formula

$$p(d_j) = \sum_{i=1}^{N} P(R_i)\ L\left[\ R_i , d_j(x_1\ x_2\ ...x_n)\ \right] \bullet$$

$$P\left(\frac{x_1\ x_2\ ...x_n}{R_i}\right) \tag{4}$$

The decision theory in its classic presentation is a one-step operation, indicating that it is based on a single calculation of the risk function based on three factors (1) matrix of a priori probabilities, (2) experimental matrix, and (3) loss-matrix.

Difficulties in the procedure (dimension of the initial data, getting a priori information and definition of loss evaluation) can be decreased using the method of "group calculation of arguments" (MGCA) (Ivakhnenko, Zaitchenko, and Dimitrov, 1976).

BASIS RULES FOR CONSTRUCTION OF MGCA ALGORITHMS

A "complete" description of the object

$$\zeta = f(x_1, x_2, ..., x_N)$$

should be replaced by some "individual" ones

$$y_1 = f(x_1, x_2) ; \quad y_2 = f(x_1, x_3) ; \quad \bullet \bullet \bullet ; \quad y_M = f(x_{N-1}, x_N) ,$$

where $M = C_N^2$;

$$\xi_1 = f(y_1, y_2) ; \quad \xi_2 = f(y_2, y_3) ; \quad \bullet \bullet \bullet ; \quad \xi_p = f(y_{M-1}, y_M) ,$$

where $p = C_M^2$ and so on.

Here, we describe a MGCA algorithm satisfying the two following conditions:

(1) Function f is the same in all the equations. Excluding intermediate variables one can get "an analog" of the complete description.

(2) The analog in its aspect must correspond to the complete description. Comparing the analog and real complete description in its common aspect one can determine "equations for construction of complete description coefficients."

With these requirements satisfied, MGCA allows the determination of evaluations of complete description coefficients even in the situation when their number is great and the quantity of the data points is less than the number of a complete polynomial.

MGCA application to the problem of recognition

In the problem of pattern recognition a possibility of transformation to a multistep presentation is in the presence of the factor $P(x_1 x_2 ... x_n / R_i)$ in Equation (4). In the first step of selection, Equation (4) "complete" description is replaced by a system of "individual" ones for all possible pairs of arguments:

$$P_{12}(d_j) = \sum_i P(R_i) L\left[R_i , d_j(x_1 x_2) \right] P\left(\frac{x_1 x_2}{R_i} \right) ,$$

$$P_{13}(d_j) = \sum_i P(R_i) L\left[R_i , d_j(x_1 x_3) \right] P\left(\frac{x_1 x_3}{R_i} \right) ,$$

$$\bullet \quad \bullet \quad \bullet \quad \bullet \quad \bullet \quad \bullet \quad \bullet \quad \bullet \quad \bullet$$

$$P_{N-1,N}(d_j) = \sum_i P(R_i) L\left[R_i , d_j(x_{N-1} x_N) \right] P\left(\frac{x_{N-1} x_N}{R_i} \right) .$$

The number of the class where a given pair is met most frequently (that is, has maximum probability of appearance) is determined for each pair of discrete variables. The number of the class determined is the decision taken with minimum risk, for example

$$y_1 = \arg \min_{d_j} p_{12}^1(d_j) = \arg \max_{R_i} P\left(\frac{x_1=0,\ x_2=1}{R_i} \right) = 1$$

indicates that a pair of discrete variables $x_1=0$, $x_2=1$ has maximum probability in the first class, resulting in taking a decision on allocating this pair to the first class. Decisions obtained for combinations of discrete variables of the first step are used as variables of the second step y_1, y_2, ..., y_p, $p = C_N^2$. In fact, passing to the second step, 1 is placed for all the pairs of discrete variables which are most frequently met in the first class; 2 is placed for all the pairs that are met in the second class most frequently, and so on. Hence, the variety of values of the second step variables is reduced abruptly.

Because at the transition from one step to another the same transformations (replacement of pairs of variables by their solutions according to minimum risk) is multiple iterated, the process of reduction of variety takes place in each step until the step s, where values of variables of the j-th representation coincide, is reached. It affords to stabilize also the solutions based on these variables, that is, for all the steps $k \geq s$ all the j-th representations will be the same.

Theorem of convergence of multistep algorithm
to the optimal decision

If x_1, x_2, ..., x_N are discrete arbitrary variables; $P(x_1, ..., x_N/R_i)$ is unknown probable distribution of the representation x_1, x_2, ..., x_N in the class R_i;

$$\arg \min_j p(d_j)$$

is given decision on the class, which representations x_1, x_2, ..., x_N according to minimum risk should be allocated to; α^k is a generalized symbol for a discrete variable of the k-th step; $\alpha^{k=1} = x$, $\alpha^{k=2} = y$, $\alpha^{k=3} = \xi$, ... ; $P(\alpha_p^k \, \alpha_M^k / R_i)$ is the probable distribution of the pair of discrete variables of the k-th step $\alpha_p \alpha_M$;

$$\arg \min_j p^k(d_j) = \arg \min_j \left\{ \sum_i P(R_i) \, L\left[R_i , d(\alpha_p^k \, \alpha_M^k)\right] P\left(\frac{\alpha_p^k \, \alpha_M^k}{R_i}\right)\right\}$$

is a decision on the class, which the pair of discrete variables of
the k-th step according to minimum risk is allocated to, then
there is such a number of step S that for all $K \geq S$ the following
statement is just:

$$\arg \min_{j} p^{k}(d_j) = \arg \min p(d_j) \ .$$

The last expression indicates that beginning with a certain step of
selection $K \geq S$, the decision on the class which any pair of
discrete variables of the k-th step is allocated to, is stabilized, that
is, it does not change with the increase of the number of k steps,
and the stable obtained solution coincides with the optimal
decision got according to the probable distribution of all variables
of the first step in accordance with minimum risk.

Description of the program

To solve the problem of pattern recognition using MGCA, a
program was compiled based on the program described by Patrati
(1971). A flowchart was added (Fig. 1), which using the matrix of
solutions constructed in every step of selection, allocates the
element of the set of studied data to one of the classes
considered.

In the flowchart of the program given here

    i -   is the number of the selection step;

    iR -   is the given number of selection steps;

    j -   is the number of the element of the set of data studied;

    jN -   is the quantity of elements of the set of data studied.

Example of solution of the recognition problem using MGCA

The methodology has been tested on the "pereriv" suite of the
Bakhar deposit, which is a complex object for interpretation of

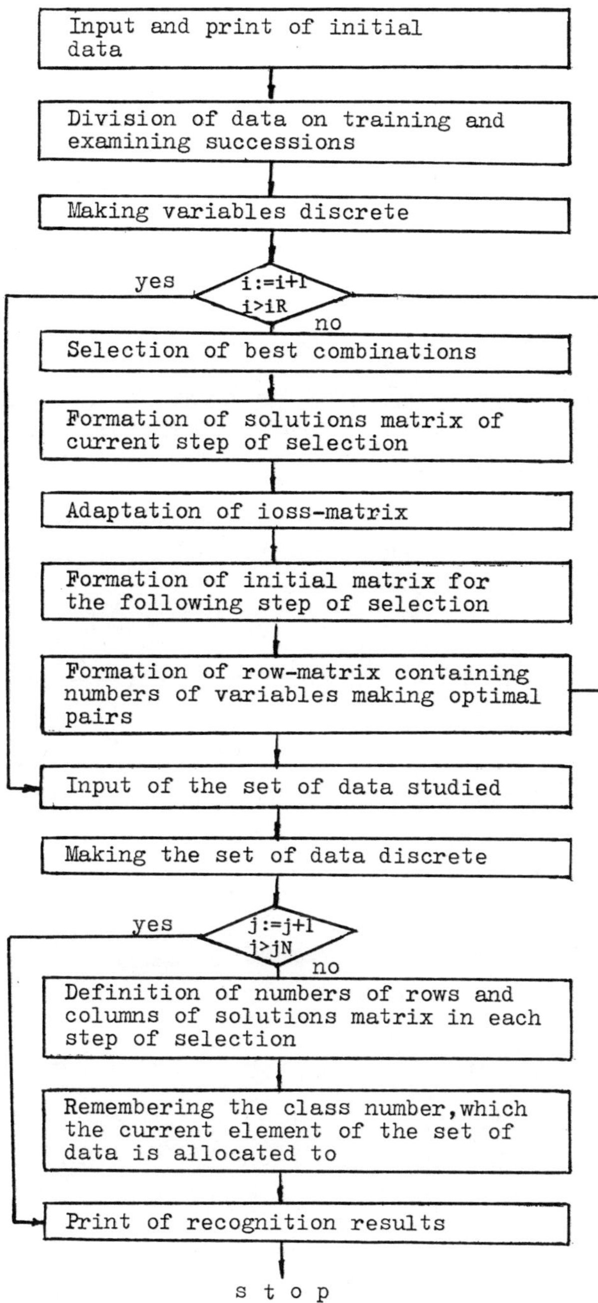

Figure 1: Flowchart of computer program for pattern recognition using MCGA.

the data on geophysical investigations of wells (GIW) and where routine ways of releasing oil- and gas-bearing reservoirs used in terrigeneous reservoirs do not give a simple solution.

The "pereriv" suite in the section of the upper part of the field productive series is characterized as mostly sandy with the content of sand and clay 13% and 27% respectively; sandstones compose 10% of sands.

From the wells having been tested 60 beds were selected (Tables 1, 2, and 3) where on each bed readings in all the parameters included into the GIW complex were taken. They are apparent formation resistivity registered by five logging sondes from the lateral electric logging (LEL), real diameter, indications of spontaneous logging, gamma-ray logging, and neutron gamma-ray logging.

To bring the parameters to the uniform conditions Tables 1, 2, and 3 give their relative values, namely :

x_1 - is the real diameter/nominal diameter ratio (d_f/d_N);

x_2 - is the relative parameter of its own polarization (ΔU_{sp});

x_3 - is the relative natural radioactivity according to gamma-ray logging $(\Delta \xi_\alpha)$;

x_4 - is the relative neutron gamma activity according to neutron gamma-ray logging $(\Delta \xi_{N\alpha})$;

$x_5 \div x_g$ is a relative parameter as a ratio of apparent electric formation resistance to mud filtrate resistance (ρ_k/ρ_f).

These parameters represent a standard GIW complex used in offshore fields in the Caspian sea.

For training the program, the selected 60 beds are divided into three classes with 20 beds in each. Beds of productive reservoirs are allocated to the first class; beds of nonproductive (aqueous) units to the second, and nonreservoirs to the third one.

Table 1. Beds used for training. Class of productive reservoirs.

$N\!\!\underline{\circ}$	d_p/d_N	Δu_{cn}	ΔJ_γ	$\Delta J_{n\gamma}$	$\rho_к/\rho_f$				
					0.5	1.05	2.25	4.25	8.25
	1	2	3	4	5	6	7	8	9
1	1.05	56	15	63	79	133	225	208	133
2	1.05	56	10	94	79	167	333	417	292
3	0.89	40	9	60	29	147	235	265	118
4	0.84	42	13	61	32	141	235	224	118
5	0.89	26	13	61	32	147	224	235	165
6	1.08	30	37	46	71	104	167	108	67
7	1.03	20	12	66	67	117	175	167	142
8	1.06	0	18	61	75	138	250	333	225
9	1.04	16	56	55	62	121	200	192	125
10	0.89	28	47	61	29	147	265	265	176
11	0.84	31	40	61	32	147	224	235	165
12	1.05	53	6	80	74	257	486	571	357
13	1.06	50	24	47	71	243	429	571	357
14	1.07	53	24	52	71	257	571	600	457
15	1.05	53	12	58	71	271	571	571	429
16	1.06	58	16	65	77	271	529	643	414
17	1.00	66	21	55	68	231	412	412	400
18	1.00	46	12	74	80	312	688	812	625
19	1.07	30	6	60	71	257	600	786	429
20	1.08	30	4	100	83	329	814	1429	1000

Table 2. Beds used for training. Class of nonproductive (aqueous) reservoirs.

№	d_f/d_N	Δu_{SP}	$\Delta \mathcal{I}_\gamma$	$\Delta \mathcal{I}_{n\gamma}$	ρ_κ/ρ_f				
					0.5	1.05	2.25	4.25	8.25
	1	2	3	4	5	6	7	8	9
1	0.97	11	3	100	57	104	150	86	39
2	1.00	20	21	90	64	107	179	157	93
3	1.00	20	21	92	71	129	214	214	121
4	0.99	20	21	92	71	121	150	264	157
5	0.99	17	14	83	71	129	221	221	136
6	1.00	20	17	92	57	71	114	193	89
7	1.00	37	31	92	64	100	136	115	82
8	1.00	33	36	88	57	107	150	129	86
9	1.00	30	29	93	71	129	193	179	129
10	0.89	20	47	65	29	94	100	106	29
11	0.98	16	13	65	50	93	121	64	43
12	0.96	11	11	60	50	79	93	50	29
13	0.97	8	13	70	50	89	121	75	25
14	1.00	16	16	50	50	86	114	75	61
15	0.97	5	33	45	50	93	146	86	64
16	0.97	6	33	50	57	100	121	86	43
17	1.13	0	41	57	50	86	136	143	79
18	0.95	8	5	50	45	71	75	43	21
19	0.97	10	0	45	50	89	114	71	54
20	0.96	25	7	45	43	64	86	71	29

Table 3. Beds used for training. Class of nonreservoirs.

N_2	d_f/d_N	Δu_{sp}	ΔJ_γ	$\Delta J_{n\gamma}$	ρ_K/ρ_f				
					0.5	1.05	2.25	4.25	8.25
	1	2	3	4	5	6	7	8	9
1	1.04	100	46	50	77	209	364	636	545
2	1.02	92	45	39	109	273	454	636	364
3	1.00	65	33	55	65	212	400	500	562
4	1.00	58	33	53	72	262	562	625	688
5	1.00	76	71	49	62	212	500	562	438
6	1.07	58	59	47	60	214	429	514	314
7	1.12	35	33	52	13	106	412	471	441
8	1.03	68	37	39	21	125	294	353	450
9	1.00	70	38	32	56	138	225	275	325
10	1.12	85	40	45	13	147	500	676	824
11	1.12	62	53	45	18	141	412	559	647
12	1.07	100	38	52	58	177	523	523	208
13	1.00	92	77	26	31	65	231	269	223
14	1.00	83	91	2	28	62	108	112	92
15	1.00	67	83	9	36	58	92	92	62
16	1.46	100	88	0	42	83	167	192	133
17	1.33	53	81	17	54	104	167	133	75
18	0.93	75	87	0	13	153	71	88	59
19	0.93	100	100	2	14	18	41	71	41
20	1.30	95	63	0	43	71	107	71	46

It is necessary to determine such a decision rule, which according to the set of parameters would allocate the current observation to one of the three classes.

The set of the data is represented as follows:

$$x_{1,1}, ..., x_{1,20}, \qquad x_{1,21}, ..., x_{1,40}, \qquad x_{1,41}, ..., x_{1,60}$$
$$x_{2,1}, ..., x_{2,20}, \qquad x_{2,21}, ..., x_{2,40}, \qquad x_{2,41}, ..., x_{2,60}$$

$$... \quad ... \quad ... \quad ... \qquad\qquad ... \quad ... \quad ... \quad ... \qquad\qquad ... \quad ... \quad ... \quad ...$$

$$x_{g,1}, ..., x_{g,20} \qquad\qquad x_{g,21}, ..., x_{g,40} \qquad\qquad x_{g,41}, ..., x_{g,60}$$

$$\text{I class} \qquad\qquad\qquad \text{II class} \qquad\qquad\qquad \text{III class}$$

The loss-matrix is selected like this:

$$
\mathbf{L} = \begin{bmatrix} 0 & 2 & 3 \\ 2 & 0 & 1 \\ 3 & 1 & 0 \end{bmatrix}
$$

The solution begins with an equal division of the data according to each class into training and examining successions. With this aim, each class dispersions are calculated in accordance with the columns; data are placed according to dispersions diminution and sorted in the following manner: columns with odd numbers are allocated to the training succession and columns with even numbers to the examining one.

The next stage is making the input variables discrete for three levels. Values of each variable were regulated according to their increase, values of intervals between the adjacent values were calculated and the value for the average interval was determined. If the given value of a variable differed from the previous one by a value more than the one-third interval, this value was put into

correspondence with the level of discretion for a unit more than
the level of discretion corresponding to the previous value; in
other situations the same discretion level was retained. Thus,
after making variables discrete for three levels the given set of
data will consist of ones, twos, and threes only.

The algorithm works with each pair of arguments successively. If
the number of features (of the initial parameters) is M = g, there
will be d = C_g^2 = 36 different pairs on the whole. As every
variable can take three values all possible combinations of values
for a pair of variables will be nine:

$$(1,1); \quad (1,2); \quad (1,3);$$
$$(2,1); \quad (2,2); \quad (2,3);$$
$$(3,1); \quad (3,2); \quad (3,3) \tag{5}$$

Experimental matrices **E** are built for training and examining
successions for each of the 36 pairs of variables taking nine
possible values.

The next matrix shows frequencies of appearance of meanings
mentioned in Expression (5) according to the classes for training
and examining successions of pairs of variables (x_1, x_3):

	Training succession			Examining succession		
	I class	II class	III class	I class	II class	III class
(1,1)	2	5	0	1	3	0
(1,2)	0	2	0	0	1	0
(1,3)	1	1	0	1	0	0
(2,1)	0	0	0	2	0	0
(2,2)	0	2	2	1	5	2
(2,3)	0	0	3	0	0	5
(3,1)	4	0	0	2	0	0
(3,2)	2	0	1	2	0	0
(3,3)	1	0	4	1	1	3

It suggests that at the training succession in the first class the pair (x_1, x_3) took values $(1,1)$ twice; meaning $(3,1)$ - four times and so on.

Furthermore, we determine the class to which according to minimum risk a current combination of meanings can be allocated. For example, for the combination $x_1 = 1$, $x_3 = 1$ of training succession, we have the following frequencies: $P_1 = 2$; $P_2 = 5$; $P_3 = 0$. Risks for allocating the pair $(x_1 = 1, x_3 = 1)$ to the j class are calculated according to the formula:

$$d_j = \sum_{i=1}^{3} L_{ij} P_i \, , \quad j = 1, 2, 3$$

Thus,

$$d_1 = 0 \bullet 2 + 2 \bullet 5 + 3 \bullet 0 = 10$$
$$d_2 = 2 \bullet 2 + 0 \bullet 5 + 1 \bullet 0 = 4$$
$$d_3 = 3 \bullet 2 + 1 \bullet 5 + 0 \bullet 0 = 11$$

Hence, according to minimum risk the pair $(x_1 = 1, x_3 = 1)$ should be allocated to the second class because

$$\min_{j} d_j = 4 \, .$$

Classes obtained for all nine combinations according to minimum risk are as follows: 2, 2, 1, 1, 2, 3, 1, 1, 3. This statement is presented as Table 4.

Table 4 was obtained according to the training succession allows to draw a conclusion that the combination $(1,1)$ belongs to the second class, combination $(1,2)$ - to the second class and so on (for the pair (x_1, x_3))

Now let us turn to the examining succession . The combination of variables $(1,1)$ in the second class at the examining succession is met thrice; combination $(1,2)$ in the second class is met once and so on (Table 5).

Table 4. Classes for all nine combinations according to minimum risk

Combinations of meanings	Number of class
1, 1	2
1, 2	2
1, 3	1
2, 1	1
2, 2	2
2, 3	3
3, 1	1
3, 2	1
3, 3	3

Table 5. Combination of variables

Combination of meanings	Number of class	Frequency
1, 1	2	3
1, 2	2	1
1, 3	1	1
2, 1	1	2
2, 2	2	5
2, 3	3	5
3, 1	1	2
3, 2	1	2
3, 3	3	3

It suggests that the number of correct decisions for the pair (x_1, x_3) at the examining succession will be:

$$3 + 1 + 1 + 2 + 5 + 5 + 2 + 2 + 3 = 24 .$$

The procedure mentioned here is iterated for all the 36 pairs of variables and for each pair at the examining succession the number of correct decisions is defined. Nine best pairs (according to the number of input features) are selected, that is those with a greater number of correct decisions, and a matrix of solutions of the first step of selection is built. Each column of this matrix contains numbers of the classes, to which the pair (x_p, x_m) can be allocated according to minimum risk when it takes the values from row (5).

	x_1x_3	x_3x_6	x_3x_7	x_3x_8	x_1x_4	x_1x_5	x_3x_5	x_3x_9	x_7x_9
1,1	2	2	2	2	2	1	1	2	2
1,2	2	1	1	1	1	2	2	1	2
1,3	1	1	1	1	2	2	1	1	1
2,1	1	2	2	2	3	3	3	2	1
2,2	2	2	3	2	3	2	2	2	1
2,3	3	1	1	1	2	3	1	3	3
3,1	1	1	1	1	3	3	3	1	1
3,2	1	3	3	3	1	3	3	3	3
3,3	3	3	3	3	1	1	3	3	1

$$(6)$$

Matrix of decisions of the first
step of selection; variables x

The multistep algorithm of taking decisions provides for the change of the loss-matrix, **L**. The elements of each column d_j of the matrix **L** are multiplied after every step by relative accuracy of recognition (for correct recognitions) in the j-th class. The result of such training of the loss-matrix is the acceleration of the convergence process of the multistep algorithm to optimal decisions. Adaptation of the matrix **L** is stopped when solutions unchanged at the transition to the next step are achieved. The constant matrix obtained is just the matrix of an optimal type because it corresponds to the maximal accuracy of solutions according to the classes. In our situation the corrected loss-matrix after the first step will be as follows

$$\mathbf{L} = \begin{bmatrix} 0 & 16.44 & 24 \\ 13.33 & 0 & 8 \\ 20.0 & 8.22 & 0 \end{bmatrix}$$

Formation of the initial matrix for the next step of selection finishes the first iteration. If in the first step of selection our variables were

$$x_1, x_2, ..., x_g$$

in the second step the variables will be as follows

$$
\begin{array}{lll}
y_1 = (x_1, x_3); & y_2 = (x_3, x_6); & y_3 = (x_3, x_7); \\
y_4 = (x_3, x_8); & y_5 = (x_1, x_4); & y_6 = (x_1, x_5); \\
y_7 = (x_3, x_5); & y_8 = (x_3, x_9); & y_9 = (x_7, x_9).
\end{array}
$$

A new matrix is achieved by substitution of combinations of values of variables (x_p, x_m) for the number of the class according to minimum risk. For example, if variables x_1, x_3 in the initial matrix had values

$$
\begin{aligned}
x_1 &: 1, 3, ... \\
x_3 &: 2, 1, ...
\end{aligned}
$$

the variable y_1, will have values

$$y_1 : 2, 1, ...$$

obtained from the matrix of solutions.

The number of selection steps R is defined from the condition:

$$R \geq \log_2 M ,$$

M - is the number of variables. The following matrices of solutions were obtained for four steps of selection.

	Y_1Y_2	Y_1Y_3	Y_1Y_4	Y_1Y_8	Y_2Y_4	Y_2Y_8	Y_3Y_7	Y_4Y_7	Y_4Y_8
1,1	1	1	1	1	1	1	1	1	1
1,2	1	1	1	1	1	1	1	3	1
1,3	1	1	1	1	1	1	2	1	1
2,1	1	1	1	1	1	1	2	2	1
2,2	2	2	2	2	2	2	2	2	2
2,3	1	3	1	3	1	3	1	3	3
3,1	1	1	1	1	1	1	1	1	1
3,2	1	1	1	1	1	1	3	1	1
3,3	3	3	3	3	3	3	3	3	3

Matrix of solutions of second
step of selection; variables Y

	Z_4Z_8	Z_1Z_3	Z_1Z_5	Z_1Z_8	Z_1Z_9	Z_2Z_8	Z_3Z_4	Z_3Z_5	Z_3Z_6
1,1	1	1	1	1	1	1	1	1	1
1,2	1	1	1	1	1	1	1	1	1
1,3	1	1	1	3	1	1	3	1	1
2,1	1	1	1	1	1	1	1	1	1
2,2	2	2	2	2	2	2	2	2	2
2,3	3	1	1	3	1	1	1	1	1
3,1	1	1	1	1	1	1	1	1	1
3,2	1	1	1	1	1	1	1	1	1
3,3	3	3	3	3	3	3	3	3	3

Matrix of solutions of third
step of selection; variables Z

	U_4U_8	U_1U_3	U_1U_5	U_1U_8	U_1U_9	U_2U_8	U_3U_4	U_3U_5	U_3U_6
1,1	1	1	1	1	1	1	1	1	1
1,2	1	1	1	1	1	1	1	1	1
1,3	1	1	1	1	1	1	1	1	1
2,1	1	1	1	1	1	1	1	1	1
2,2	2	2	2	2	2	2	2	2	2
2,3	1	1	1	1	1	1	1	1	1
3,1	3	3	1	3	1	3	3	1	1
3,2	3	3	1	3	3	3	3	3	1
3,3	3	3	3	3	3	3	3	3	3

Matrix of solutions of fourth
step of selection; variables U

The second stage of the program solves the problem of
recognition for the elements of the input set of data using
matrices of solutions built at the first stage. The set of data is
subjected to making discrete for three levels. The levels were
defined by the first stage of the program. For example, let an
element of the set of data have the following set of meanings of
the features

$$2, 2, 2, 2, 2, 3, 2, 2, 2$$

One should define the class to which it is allocated.

The best decisions always are in the first row of the matrix of
solutions. In our situation it is (U_4, U_8). Here is the solution
made by the program:

$$d(U_4, U_8) = 1$$
$$d(U_4) = d(Z_1, Z_8) = d(1, 2) = 1$$
$$d(U_8) = d(Z_3, Z_5) = d(2, 1) = 1$$
$$d(Z_1) = d(Y_1, Y_2) = d(2, 1) = 1$$
$$d(Z_8) = d(Y_4, Y_7) = d(2, 2) = 2$$
$$d(Z_3) = d(Y_1, Y_4) = d(2, 2) = 2$$
$$d(Z_5) = d(Y_2, Y_4) = d(1, 2) = 1$$
$$d(Y_1) = d(x_1, X_3) = d(2, 2) = 2$$
$$d(Y_2) = d(X_3, X_6) = d(2, 3) = 1$$
$$d(Y_4) = d(X_3, X_8) = d(2, 2) = 2$$

Table 6. Beds for recognition

N	α_f/α_N	Δu_{sp}	ΔJ_γ	$\Delta J_{n\gamma}$	ρ_k/ρ_f				
					0.5	1.05	2.25	4.25	8.50
	1	2	3	4	5	6	7	8	9
			Productive reservoirs						
1	1.08	35	22	61	108	208	333	375	225
2	1.04	56	6	55	67	125	200	167	117
3	1.12	35	33	52	13	106	412	471	441
4	0.98	15	46	50	29	129	176	182	135
5	0.93	42	26	59	29	147	265	206	106
6	0.93	37	16	86	31	165	353	324	188
7	0.89	35	13	100	29	200	471	500	382
8	1.00	5	17	47	70	225	388	312	200
9	1.00	25	10	74	69	225	375	300	275
10	1.00	45	17	58	62	212	312	200	200
11	1.00	56	15	56	69	225	425	275	212
12	1.00	43	13	70	72	250	412	338	312
13	1.00	65	33	55	65	212	400	500	562
14	1.00	39	8	58	70	238	438	425	400
15	1.00	58	33	53	72	262	562	625	688
16	1.00	58	17	49	72	225	425	438	438
17	1.06	0	6	65	76	271	500	600	371
18	1.07	42	18	52	71	229	471	543	314
19	1.08	20	12	60	77	271	643	857	543
20	1.08	50	18	52	66	243	457	500	357
21	1.08	30	4	100	83	329	814	1429	1000
22	1.04	53	14	55	71	243	486	600	386
23	1.04	53	26	60	71	243	471	529	471

PATTERN RECOGNITION IN GEOLOGY   275

Table 6. continued.

N°	d_f/d_N	Δu_{sp}	$\Delta \mathcal{I}_\gamma$	$\Delta \mathcal{I}_{n\gamma}$	ρ_k/ρ_f				
					0.5	1.05	2.25	4.25	8.50
	1	2	3	4	5	6	7	8	9

Nonproductive (aqueous) reservoirs

N°	d_f/d_N	Δu_{sp}	$\Delta \mathcal{I}_\gamma$	$\Delta \mathcal{I}_{n\gamma}$	0.5	1.05	2.25	4.25	8.50
24	0.97	11	14	50	54	100	150	129	64
25	0.97	23	16	45	46	79	93	61	43
26	0.97	6	22	75	50	82	114	72	32
27	0.97	0	16	65	50	79	93	46	29
28	0.89	3	23	65	29	147	218	118	76
29	1.00	27	26	66	57	100	143	136	79

Nonreservoirs

N°	d_f/d_N	Δu_{sp}	$\Delta \mathcal{I}_\gamma$	$\Delta \mathcal{I}_{n\gamma}$	0.5	1.05	2.25	4.25	8.50
30	1.07	100	38	52	58	177	523	523	208
31	1.12	77	60	24	15	176	324	353	382
32	1.00	54	20	62	35	176	294	382	265
33	1.02	54	46	65	100	218	427	500	273

Table 7. Recognition results

| Recognition results | | |
1 class	2 class	3 class
1	24	5
2	26	6
3	27	13
4	29	25
7	–	28
8	–	30
9	–	32
10	–	33
11	–	–
12	–	–
14	–	–
15	–	–
16	–	–
17	–	–
18	–	–
19	–	–
20	–	–
22	–	–
23	–	–
31	–	–

1,2,3... etc. are numbers of beds according to the order from
Table 6.

$$d(Y_7) = d(X_3, X_5) = d(2, 2) = 2$$
$$d(Y_1) = d(X_1, X_3) = d(2, 2) = 2$$
$$d(Y_4) = d(X_3, X_8) = d(2, 2) = 2$$
$$d(Y_2) = d(X_3, X_6) = d(2, 3) = 1$$
$$d(Y_4) = d(X_3, X_8) = d(2, 2) = 2$$

To examine the quality of algorithm recognition on the base of the tested wells not included into training we selected a control set of data from 33 beds, consisting of 25 beds of the first class; 6 beds of the second class and four beds of the third class. These beds with characteristics according to the GIW complex are given in Table 6 with a through numeration.

As it follows from Table 7 showing the results of recognition the error was made in six situations only.

The algorithm and program developed with use of decision theory can be recommended for wide application in solving the problem of pattern recognition in different spheres in geology.

REFERENCES

Ivakhnenko, A. G., Zaitchenko, Yu. P., and Dimitrov, V. D., 1976, Taking decisions on the base of self-organization: M., "Sovetskoye radio".

Patrati, I. Z., 1971, Program in the language of ALGOL-BESM-6 translator for solving the problem of recognition patterns (classes) according to algorithms of decision theory: Avtomatika, no. 3, p. 83-85.

Truxal, J. G., and Padallino, J., 1961, Theory of decisions, in Mishkin, E., and Braun, L., eds., Adaptive control systems: McGraw-Hill Book Co., New York, p. 479-501.

LITHOLOGICAL-MINERALOGICAL PECULIARITIES OF SEDIMENTARY ROCKS IN ABNORMAL THERMOBARIC CONDITIONS AND PREDICTION OF OIL AND GAS CONTENT AT GREAT DEPTHS

R. D. Djevanshir

Azerbaijan Academy of Sciences

ABSTRACT

Experimental investigations of Pliocene clays in the Baku Archipelago have noted that the persistence of montmorillonite content to 6200 m is connected with predominance of the process of secondary montmorillonite formation from illite over the process of primary montmorillonite destruction. Based on the mechanism of interrelation and reciprocal influence of clayey-mineral transformation and thermobaric conditions of the sedimentary basin, the author has developed a method for computer prediction of filtration-capacity parameters for terrigenous sedimentary rocks taking into account the influence of thermobaric, lithological-mineralogical, structural, and geotectonic factors. For zones of abnormally high pore pressure and temperature decline, the Southern Caspian Depression (Basin), is a good example where the possibility of montmorillonite preservation to 15-17 km may indicate the probability for discovery of commercial hydrocarbon accumulations at depths of 9 km.

INTRODUCTION

From the position of the 'systems approach', oil and gas content of a sedimentary formation is a function of reservoir properties, condition of organic matter transformation, and hydrocarbon

migration. Evaluation of oil and gas content at great depths when
direct geological, geophysical, and geochemical data are absent,
can be realized on the basis of a forecast, keeping in mind, not a
formal extrapolation of data, but a prediction with consideration of
the reciprocal influence of lithological-mineralogical and
thermobaric factors. These problems were studied on an example
of a Cenozoic sedimentary sequence of the Southern Caspian
Depression which is characterized by high-sedimentation rates (to
1300 m/mln years,), an enormous (to 25 km) thickness of the
sediments, abnormally high pore pressures (average gradient is
0.018 MPa/m), and a lowered geothermal field (average
geothermal gradient is 16°C/km). Special attention was paid to
studying clays and clayey rocks composing from 50 to 95% of the
sandy-aleurite-clayey sedimentary sequence and playing a leading
role in forming lithological-mineralogical sequence and
thermobaric characteristics of the formation considered.

CONTENT

Transformation of montmorillonite and illite-montmorillonite
minerals into illite under action of diagenesis and katagenesis is
described for many large sedimentary basins in the world. It
should be noted that these modifications of clayey minerals in
katagenesis are obligatory (and not simply possible as in
diagenesis), that is they are connected with the thermobaric
parameters which increase during burial of sediments. As a result
in late katagenesis, associations of clayey minerals become
two-components consisting of illite and chlorite, independent of
their original composition.

In spite of this relationship, discoveries were reported in recent
years of essentially unchanged montmorillonite at great depths
and in large quantities (Belov and others, 1978; Zkhus and
Bakhtin, 1979; Kheirov, 1979). According to Zkhus and Bakhtin
(1979) this fact does not contradict the existing concepts about
stage transformations of clayey minerals. In particular, they
connect montmorillonite preservation at depth on the order of
5500 m in sequences of the Fergana Depression with its
volocanogenosedimentary origin and to the absence of organic
matter in rocks. Belov and others (1978) explain the discovery of
almost unchanged montmorillonite in Pliocene sediments of the
Baku Archipelago at 6026 m to the result of specific
sedimentological conditions, composition of the original matter,
and influence of abnormally low temperatures. Such an

explanation sounds well founded enough, but at the same time, the greatest interest is the identification of montmorillonite at considerable depth. The study of regularities of clayey-mineral distribution through the entire sequence, exposes major reasons for influencing deviations in clayey-mineral transformation, and prediction of katagenetic transformations at depths not penetrated by drilling wells are of great importance.

In order to study the character of distribution of clayey minerals with depth in the productive series of the Baku Archipelago, X-ray-structural[1] and scanning-electron microscope (SEM) investigations of about 140 clayey and clayey-rock samples taken from wells to 6200 m (Table 1) were made. The data of the X-ray-structural investigations obtained previously by Kheirov (1979) were used as well.

The basic mass of clayey minerals in rocks of the productive series consists of the montmorillonite (average 40%) and illite (37%) groups. Kaolinite content ranges from 15 to 20%, chlorite from 5 to 10%, and mixed-layered minerals from traces to 5%. According to the data of the granulometric analysis of rocks, a fraction of size under 0.001 mm shares 68.8% on the average. Sandy-aleurite fractions average 22% and carbonates 15%. Porosity of clays at 1000-6000 m ranges on the average between 28 to 10%; permeability of the calculated data makes up (0.6-250) 10^{-6} fm^2 (Table 1) (Buryakovsky, Djevanshir, and Aliyarov, 1983).

Charts showing dependencies of clayey-mineral content versus depth were constructed according to the X-ray-structural analysis data (Fig. 1). As we see, an obvious regularity in changing clayey-mineral content in the sequence of the Baku Archipelago is not observed. Montmorillonite in large quantities (40% and more) is present for the entire depth interval.

Hence, one can draw a conclusion that in clays of the productive series to 6200 m at least, evidence of montmorillonite transformation to illite is not present. Kheirov (1979) drew the same conclusion considering that the geotectonic conditions of

---

[1]X-ray-structural analysis was carried out in AzNIPIneft under the guidance of M.B. Kheirov.

Table 1. Characteristics of clays and clayey rocks of productive series of Baku Archipelago.

Depth intervals, m	Clay, materials content, percent					Porosity, percent	Permea-bility, 10^{-9} darcies	Pore diameter, microns	Pore pressure gradiant, megaPascal per m	Geo-thermal gradient, °C per km
	mont-morillonite	illite	kaolinite	chlorite	mixed-layered					
1	2	3	4	5	6	7	8	9	10	11
1000 - 2000										
average	32.5	43.5	17.5	6.5	traces	28	142	2.7	0.016	12
range	10 - 45	35 - 65	15 - 20	5 - 10		22 - 33	35 - 250	1.7 - 3.6	0.012 - 0.020	10 - 15
2000-3000										
average	45.0	35.0	13.0	7.0	traces	20	22	2.1	0.017	11
range	35 - 70	20 - 40	0 - 15	0 - 10		15 - 28	8 - 35	1.3 -3.1	0.013 - 0.021	10 -12
3000-4000										
average	36.0	42.0	14.0	7.0	1.0	17	6	1.6	0.018	10
range	15 - 50	30 - 60	5 - 20	5 - 15	0 - 5	8 - 24	3 - 8	1.0 -2.5	0.014 - 0.022	8 - 11
4000-5000										
average	40.0	38.0	12.5	5.5	4.0	14	2	1.3	0.019	17
range	15 - 70	10 - 60	0 - 20	0 - 10	0 - 30	2 - 21	1 5 - 3	0.7 - 2.0	0.015 - 0.023	15 - 19
5000-6000										
average	39.0	39.0	15.5	5.0	1.5	12	1	0.8	0.019	22
range	5 - 65	20 - 65	0 - 30	0 - 15	0 - 15	0 - 18	0.8 - 1.5	0.5 - 1.5	0.015 - 0.023	21 - 23
over 6000										
average	36.0	37.5	15.5	4.0	7.0	10	0.7	—	0.020	20
range	5 - 70	20 - 60	10 - 25	0 - 10	0 - 25	0 - 16	0.6 - 0.8		0.016 - 0.024	15 - 25

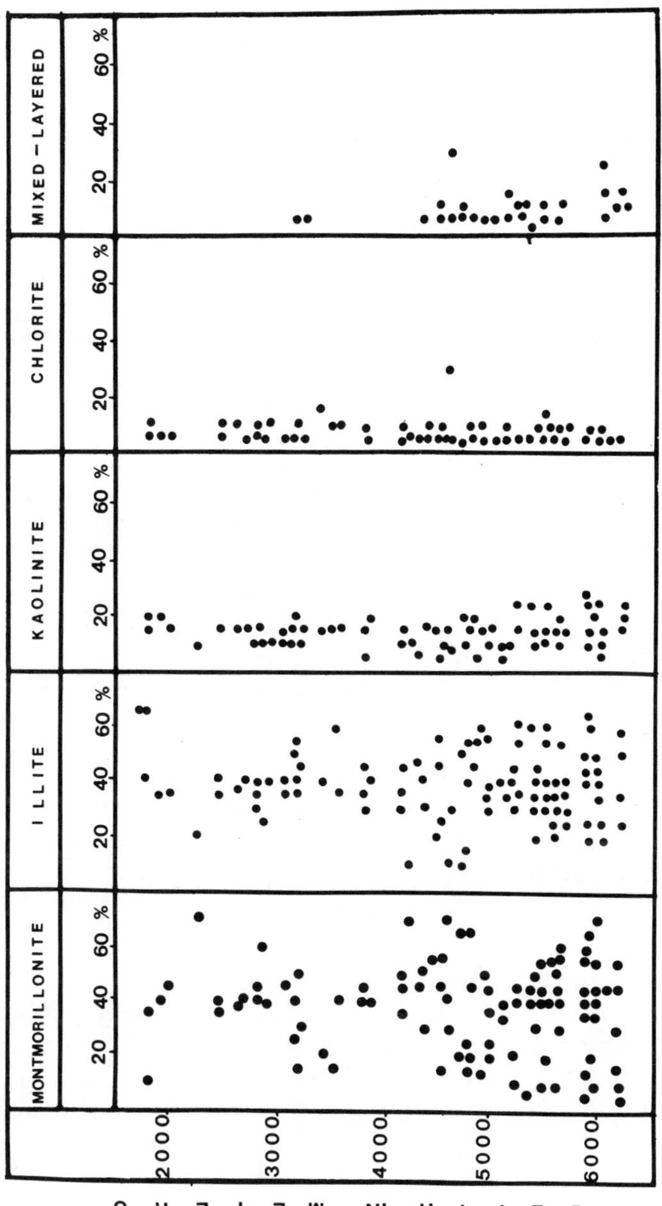

Figure 1. Clayey-mineral content in rocks of productive series of the Baku Archipelagc.

the basin and permeability of rocks are major factors influencing transformation of clayey minerals.

The absence of clear traces of clayey-mineral transformation in the region studied can be explained also by abnormally low temperatures. The works by Khitarov and Pugin (1966) and Powers (1967) have shown that temperature is a principal factor influencing the process of montmorillonite dehydration. In the sequence of the Baku Archipelago an average value of the geothermal gradient is approximately 16°C/km, and at depths on the order 6000 m, the temperature does not exceed 110°C. Of course, such low thermal regime of the medium does not contribute to processes of montmorillonite transformation into illite. Nevertheless, it should be noted that this factor should have lowered the rate of montmorillonite dehydration only. The process itself of transformation of this mineral with depth at other equal conditions must have taken place. All the mentioned reasons allow an assumption that the process of transformation of clayey-minerals is influenced by some other factors, the action of which have conditioned the picture fixed by the X-ray-structural analysis.

One should notice another important detail making interpretation of the results of the X-ray-structural analysis of clayey rocks difficult. The thing is that these results do not always allow a judgment about the origin of clayey minerals, that is if they are primary or secondary through the entire depth interval studied. For example, Millot (1968) notes that montmorillonite formed at the final stage of illite dehydration does not seem to differ much from real montmorillonite with which it has the same X-ray-graphical characteristics.

The results of the analyses of about 200 SEM photographs of surfaces of clay-chip samples are of interest (Fig. 2). The surface of clayey-rock samples was studied parallel, perpendicular, and oblique to the bedding. The analysis of the microphotographs affirms that in a quantitative correlation the mineralogical composition of the Baku Archipelago clays is preserved on the whole, through the depth interval studied. The major clayey minerals are illite and montmorillonite; kaolinite and chlorite are subordinate. The rocks have a close to cellular (honeycomb) texture that is well seen in oblique chips.

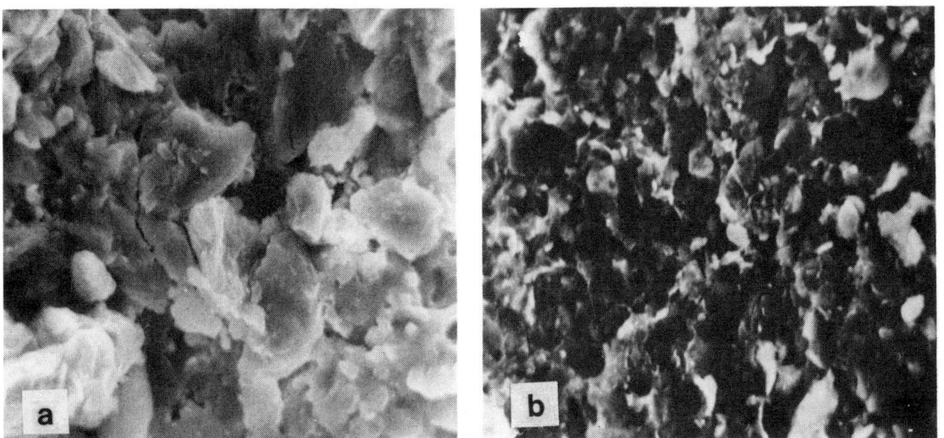

Figure 2. Microphotographs of surfaces of clay-sample chip, x 1000.

An interesting and new result of SEM interpretation notes the presence of clayey-mineral transformation both in straight and reverse directions; the processes takes place parallel to basin depressions. On the photographs of the samples from the interval of 1400-1600 m changes of clayey minerals in their mass are rather small although one can distinguish destruction of separate layers, their swinging to circumference (periphery), presence of secondary pores and fracturing in separate grains of illite. There also are microcaverns formed due to secondary processes. The results of clayey-mineral transformation are shown vividly on the photographs of samples taken from the depth of 4000 m. Illite and montmorillonite prevail in the mass; as for montmorillonite, it is both, primary and secondary. The latter spreads in openings between illite developing on its periphery and in the fractures. Primary montmorillonite destructs or swings to the periphery, secondary pores being observed.

Thus, the process of illite dehydration with its transformation into montmorillonite makes an advance in the sequence of Pliocene sediments of the Baku Archipelago with depth, together with destruction and dehydration of primary montmorillonite. At first sight this paradoxical fact is conditioned apparently by the preservation of quantitative correlations between illite and montmorillonite with depth to a great extent.

It should be noted that the formation of montmorillonite by illite degradation and a possible mechanism for this phenomenon has been considered in a number of works (Millot, 1968; Klubova, 1973). Millot noted that many properties of different three-layer minerals such as an ability for swelling, particle size, and chemical composition change in the same manner. Fine particles which have the most ability to swell, are subject to leaching with natural solutions most easily. For all this, interbedded ions are removed, and then the chemical composition of particles changes to approach montmorillonite. Klubova (1973) considers both isomorphic replacements in the grid and increase of a degree of dispersion leading to formation of incompensated relations to be the most essential factors for the growth of sorptive properties of clayey minerals. Within the group of illites differences connected with their hydration leading to their successive potassium loss and its replacement with hydrogen also are important. These differences provide conditions for partial intracrystalline swelling of changed (degraded) illites. Yet it should be noted that there is no unity of thought among the researchers about the problem on the mechanism of illite degradation.

What are the reasons for the picture of clayey-mineral transformation observed in the sequence of the Baku Archipelago? It is clear that there are factors here that together with the geochemical regime of the medium on one hand, decrease the rate of montmorillonite transformation into illite considerably, and on the other, make for developing the process of illite degradation into montmorillonite. One of these factors to be considered is the abnormally low geothermal gradient. The other factor is an abnormal high pore pressure in clays developed intensively through the entire sequence of the productive series. Hydrostatic gradients of pore pressures in clays in the interval of 1000-6000 m according to the data of about 2000 determinations obtained by methods of field geophysics and drilling, range from 0.012 to 0.024 MPa/m with an average value of 0.018 MPa/m (Fig. 3 and Table 1) (Buryakovsky, Djevanshir, and Aliyarov, 1983).

It is known that the nature of abnormal pore pressures is multiform and can be defined by different natural factors which may be imposed upon each other. However, for the region of the Baku Archipelago in the areas of accumulation of thick series of terrigeneous, mainly clayey rocks, the most probable mechanism

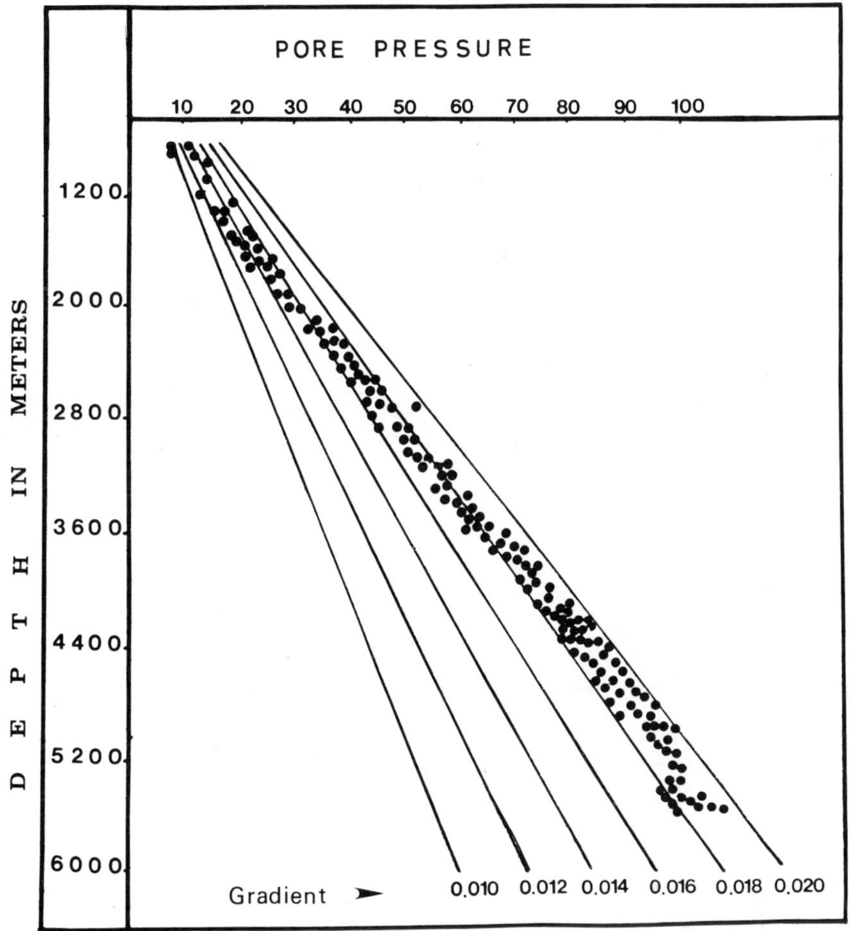

Figure 3. Pore pressure change in clays with depth in sequence of Baku Archipelago. η is pressure gradient.

for abnormal pressure development is gravitational-filtration rock consolidation. Prevailing on the factor of gravitational consolidation over a filtration component observed in conditions of high rates of sedimentation leads to considerable nonconsolidation of clayey rocks and development of abnormal pressures within them. This factor acts everywhere during the process of evolution of a sedimentary basin.

A problem of the analytical description of processes of geofluid (pore) pressure formation and destruction has been solved (Buryadovsky, Djevanshir, and Aliyarov 1983). Dynamic models satisfying the following conditions have been suggested: (1) a current geofluid pressure (a geofluid pressure at any moment of time) is a result of dynamic equilibrium between synchronously flowing processes of formation and destruction of these pressures in the given geological object; (2) natural factors influencing formation and destruction of geofluid pressures act permanently; (3) the rate of changing a geofluid pressure in the given geological object is proportional to the exiting pressure; (4) pressures increase and drop so that constant shares of the existing value of the geofluid pressure increase and decrease per unit of time (the given condition is not strictly obligatory); and (5) factors of pressure drop act so that a share of pressure drop per unit of time is proportional to the product of increase shares of the pressure by its drop shares. Dynamic models obtained are described by a system of nonlinear differential first-order equations such a

$$\frac{dp_1}{dt} = \varepsilon_1 P_1 - \gamma_{12} P_1 P_2$$

$$\frac{dp_2}{dt} = -\varepsilon_2 P_2 + \gamma_{21} P_1 P_2 \qquad (1)$$

where $P_1 = P_1(t)$ is the pore pressure during the period of its increase; $P_2 = P_2(t)$ is the pore pressure during the period of its drop; and ε_1 and ε_2 are coefficients of pore pressure change during the periods of its increase and drop, respectively; γ_{12} and γ_{21} are coefficients of interaction of factors defining preservation or change of the pressure. Equation (1) allows a description of a process of formation, stabilization, and destruction theoretically of pore pressures of any genesis. Because of the difficulty in simultaneous experimental determining the coefficients of pressure change and coefficients of reciprocal influence of the factors, numerical calculations according to the models are possible in a particular situation only when the coefficient of reciprocal influence of the factors may be neglected. For $\gamma_{12} = \gamma_{21} = O$, Equation (1) is two equations, one of which describes the increase of pore pressure to abnormally high, and the other one, its drop to conditional hydrostatic. For conditions of gravitational-filtration origin of abnormally high pore pressure, it is necessary also to take into account the effect of self-regulation of the process, that for one-stage formation of pressures which leads to the following equation

$$ p_1 = \frac{p_{max}\, p_o\, e^{\varepsilon}\, 1^{p}\, m^{ax^t}}{p_{max} - p_o\, (\, 1 - e^{\varepsilon}\, 1^{p}\, m^{ax^t}\,)} \qquad (2) $$

where P_0 is the initial value of pore pressure corresponding to hydrostatic pressure of water at a depth where sedimentation began; P_{max} is a maximum possible pore pressure close numerically to the rock pressure; and t is time. The proportionality factor calculated for the sequence of the Southern Caspian Depression averages 0.02 [MPa x mln years]$^{-1}$.

The character of pore pressure change with depth is supposed to be analogous to pressure change in time and may be described by an equation analogous to (2). This assumption may be considered likely for conditions of the Southern Caspian Basin, at least, taking into account a relatively young age of deposits, absence of noticeable structural reconstructions, one-stage formation of a folded structure, and normal bedding of sequential stratigraphic intervals among others.

Other factors also can influence the development of abnormal pressures, but in a given region, in the opened part of the sequence they play a subordinate role (Buryakovsky, Djevanshir, and Aliyarov, 1983).

The role of pore pressure in diagenesis and katagenesis of clayey minerals is considered rather insufficiently in the literature. One can show theoretically that increasing pressure lowers the rate of reaction of montmorillonite dehydration. As the reaction of clay transformation into illite includes the increase of a pure water volume due to the release of layers with connate water which are denser in comparison with free water, any factor acting against this increase of the volume (such as pore pressure) may decrease the rate of dehydration reaction. This hypothesis is in a good agreement with picture observed in the sequence of the Baku Archipelago.

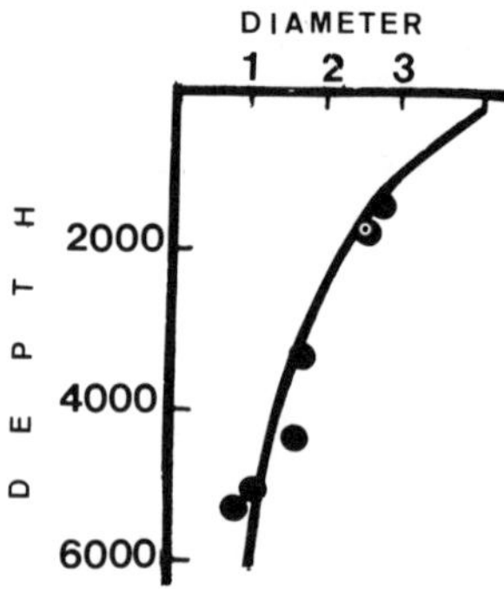

Figure 4. Change of cross size of pores with depth in clayey rocks of Baku Archipelago.

On the other hand, abnormally high pore pressure can lead to illite degradation with formation of secondary montmorillonite due to absorption of water by illite in the pore space of clays. One should note that in similar conditions, destruction of clayey minerals and increase of their dispersion makes for the process of illite degradation during basin depression. This phenomenon is observed in the sequence of Pliocene sediments of the Baku Archipelago according to the SEM data on the change of cross size of pores with depth (Table 1 and Fig. 4) (Buryakovsky and Djevanshir, 1979). As one can see in the interval of 1400-5200 m a median pore diameter decreases regularly with depth according to an exponential-type curve.

Evaluation of influence of the process of illite degradation on geothermal characteristics of the sequence is of interest. It is known (Zlochevskaya, 1976) that the reaction of hydration of clayey rocks is exothermal. At equal-conditions, other depth intervals where the process of illite degradation takes place, can be characterized by increased geothermal regimes.

In connection with this, the results of temperature measurements taken in deep wells drilled in areas of the Baku Archipelago are of interest. The work of Narimanov (1984) reveals a characteristic peculiarity of the region considered, that is an increase of the geothermal gradient with a depth of order 4 km (Table 1). Analyzing the reason for this phenomenon, Narimanov comes to the conclusion that it is likely to be in lithological-mineralogical peculiarities of the Baku Archipelago sediments. Taking into account the fact that the process of montmorillonite dehydration which plays the role of a screen on the way of a deep thermal flow to the Earth's surface. However, as it has been noticed that the real data (Table 1 and Fig. 1) do not confirm presence of a zone of mass montmorillonite dehydration in the Baku Archipelago sequence. That is why it seems impossible to explain the reason for changing the geothermal gradient from a depth on the order of 4 km in this manner.

Supposedly there are grounds to assume that the geothermal gradient growth from a depth on the order of 4 km may be connected with the process of illite degradation accompanied by thermal energy release. It seems at a depth on the order of 4 km, the rate of illite degradation exceeds a certain limit that leads to the prevailing of the process of hydration over dehydration. Therefore, for predicted evaluations the influence of diagenetic and katagenetic transformations on a thermal regime of the medium must be taken into account equally with regard for the influence of temperature on diagenetic and katagenetic processes.

An additional problem is specification of depth limits for possible montmorillonite preservation depending on thermobaric factors. In this connection, the results of experimental studies of Khitarov and Pugin (1966) which allow the evaluation of depths of montmorillonite extension in different thermobaric conditions are of great interest. So, when changing a geothermal gradient from $40^{\circ}C/km$ to $10^{\circ}C/km$ ultimate depths of montmorillonite extension range between 3 to 16 km. For the Baku Archipelago sequence in particular, where the average geothermal gradient is $16^{\circ}C/km$, the ultimate depth of montmorillonite preservation may reach 8-9 km. An analytical expression for the dependence of depth of montmorillonite extension (H, km) versus the geothermal gradient (G, $^{\circ}C/km$) is achieved according to the data of Khitarov and Pugin (1966) as follows

$$H = 261 \, G^{-1.23} \tag{3}$$

Together with Equation (3) one can state that in sedimentary sequences characterized by abnormally high pore pressures, depths of possible montmorillonite extension can be deeper. Keeping in mind a linear dependence of pressures in sedimentary basins versus depth, the expression for evaluation of ultimate depths of montmorillonite preservation (H_{max}) is

$$H_{max} = 261 \ K_a G^{-1.23} \qquad (4)$$

where K_a is a dimensionless factor of anomalous pore pressure equal to the ratio of a read (predicted) pore pressure to conditional hydrostatic. Considering Equation (4) for g=16°C/km and K_a=1.8, ultimate depths of montmorillonite extension in the Southern Caspian Depression are up to 15-17 km.

To predict oil and gas content at great depths, it is necessary to evaluate the predicted values of filtration-capacity parameters of the reservoir. With this aim, a theory, method, and technique for predicting physical properties of sedimentary rocks have been developed (Buryakovsky, Djafarov, and Djevanshir, 1982).

A notion of the lithogenetic petrophysical system functioning during the process of sediment accumulation, their consolidation and transformation into sedimentary rocks is the basis for constructing theoretical models. Hooke's Law about the proportionality of a relative volumetric change of bodies to the load applied is the experimental basis for these assumptions. When simulating parameters of rocks properties, this model may be realized both without restrictions and with a limitation of the rate of change of the parameter studied. In this dependence, the parameters change may be according either to an exponential or a logistic curve. We have calculated models porosity, permeability, and density of terrigeneous sedimentary, the basis of which is a multiplicative model of sediments consolidation. The model of porosity, in particular, is

$$m = \frac{m_o \prod_{i=1}^{n} X_i}{1 - m_o \left(1 - \prod_{i=1}^{n} X_i\right)} \qquad (5)$$

where m_0 is initial porosity; m is porosity at depth (or to a certain moment of a rock's consolidation time); and X_i is numerical evaluation of a rock's consolidation value containing coefficients X_i, each carrying some information about the value of this or that natural factor. The model of the sediment-consolidation process is a product of independent normed coefficients, each reflecting a degree and character of natural factors influence both, on the entire complex of sedimentary rocks and on each parameter studied. This model contains eight coefficients corresponding to factors of external stimulation (geologic age, dynamic and static loads, temperature, intensity of sedimentation) and lithological factors characteristic of rocks themselves (mineralogical composition of clays, degree of cementation and grading).

Simulation coefficients are dimensionless values, each depending on one natural factor or one group of interrelated factors. The use of dimensionless values is conditioned by the fact that the difference in dimensions of natural factors does not allow the possibility to link uniformly all these factors among themselves within the limits of a single model.

Numerical determination of simulation coefficients is made on the basis of physical experiments and natural measurements by specifying normal relations of numerical meanings of natural factors with the parameter studied, and by definition of a normal share of the influence of each natural factor on the parameter studied.

In order to get the results of simulation as distributions of parameters of reservoir properties, and taking into account the impossibility of a precise setting of simulation coefficients and numerical evaluations of natural factors, they are given in an interval form. Calculations of the models are realized using a Monte-Carlo method utilizing a computer program consisting of subprograms on the generation of random numbers calculated for the models suggested and statistical processing of sets of numbers obtained. Results include average evaluations, confidence intervals, coefficients of variation, and histograms of distribution of the parameters studied.

The method is used for predicting properties of reservoirs of the Southern Caspian Depression to 9000 m (Fig. 5, Fig. 6). A comparison of the calculated data with core analyses results is satisfactory.

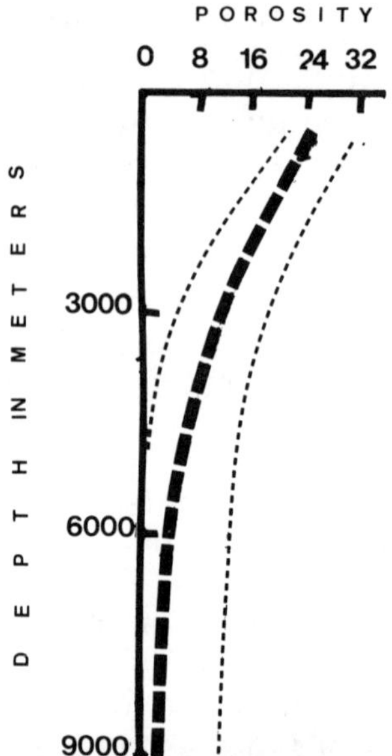

Figure 5.   Simulation results of clay porosity; 1 is average value, 2 is two-sigma limit, 3 is core data.

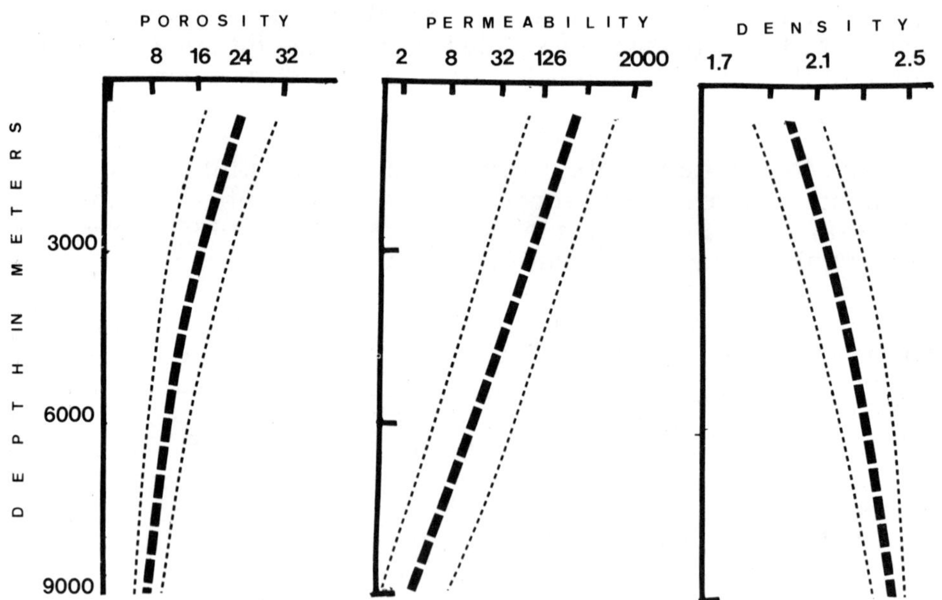

Figure 6.   Simulation results of porosity (a), permeability (b), and density (c) of reservoir; 1 is average curve.

The results allow a complex evaluation of oil and gas content of the sequence of the Southern Caspian Depression at depths not penetrated by drilled wells. Screening properties of clay caps at depths of 6.5 km will be preserved owing to the presence of a considerable quantity of montmorillonite and abnormal high pore pressures preventing fluid filtration through the cap. Porosity of clays at depths of the order of 9 km may reach 10% that may be evidence for the presence of abnormally high pore pressures of gravitational-filtration genesis at great depths. This also is included in the results on the study of porometric characteristics of clays which shows that the clays are unconsolidated because the median pore diameter in the penetrated part of the sequence does not exceed 1 mkm. Porosity of reservoirs at depths on the order of 9 km averages 7-9%, and the permeability 1.5-11 fm^2.

Presence of reliable impermeable rocks together with good filtration-capacity; parameters of reservoirs, abnormally high pore pressures, and relatively low temperatures that make impossible preservation of hydrocarbons, allow the confident prediction of the probability of discovery of commercial oil and gas accumulations in the Southern Caspian Depression at depths of 9 km plus.

The regularities pointed out are most characteristic of young areas of rapid accumulation of thick series of terrigeneous sedimentary rocks. The analysis of real curves of clayey-rock consolidation in different regions of the world (Fig. 7) shows that preservation of a high baric potential for ancient sedimentary series is problematic. In view of this the peculiarities of the difference of oil and gas content in ancient sedimentary formations at high depths from young unconsolidated sediments, which have not yet reached a high degree of diagenetic and katagenetic transformation, can be explained.

CONCLUSIONS

(1) Experimental investigations of Pliocene clays of the Baku Archipelago have stated that the observed persistence of montmorillonite content of 6200 m is connected with predominance of the process of secondary montmorillonite

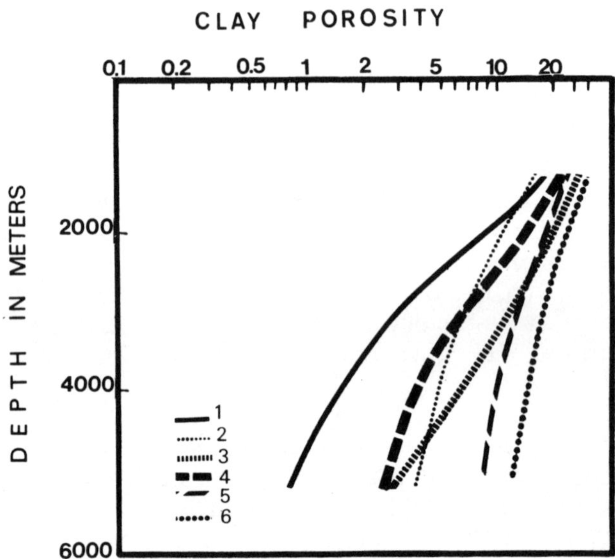

Figure 7. Dependence of clay porosity versus depth of their
           bedding; 1 is according to J. Weller (Devonian); 2 is
           according to B.K. Proshlyakov and V.M. Dobrynin
           (Mesozoic); 3 is according to N.B. Vassoevich
           (Oligocene-Miocene); 4-6 are according to A.G.
           Durmishyan and others (Middle Pliocene); 4 is for
           Apsheron Archipelago; 5 is for Southern-Apsheron
           water area zone; 6 is for Baku Archipelago and
           Prikurinskaya Lowland.

formation from illite over the process of primary montmorillonite destruction.

(2) During the process of diagenesis and katagenesis reciprocal influence of clayey-mineral transformation and thermobaric conditions of the sedimentary basin takes place. The process of montmorillonite dehydration with its transformation into illite develops under the influence of temperature increase with depth, and together with this process abnormal high pore pressures may form in clays. On the other hand, an abnormally high pore pressure (of any genesis) prevents the process of montmorillonite dehydration and makes for development of illite degradation with the formation of secondary montmorillonite. Illite degradation similar to any exothermal process is accompanied by heat release influencing the geothermal characteristic of the sequence. Such phenomena are most characteristic of young areas of rapid accumulation of thick series of terrigeneous mainly, clayey sedimentary rocks. The Southern Caspian Depression is an example of such a region.

(3) The experimental materials obtained with regard for the data (Khitarov and Pugin, 1966), suggest a calculation formula for evaluation of the ultimate depth of possible montmorillonite extension for any thermobaric conditions including the condition when the real pore pressure in clays differs from the conditional hydrostatic one. In the sedimentary sequence of the Southern Caspian Depression calculated values for ultimate depths of montmorillonite extension with regard for the predicted values of thermobaric parameters are 15-17 km.

(4) A method for computer prediction of filtration-capacity parameters of terrigeneous sedimentary rocks with consideration of the influence of thermobaric, lithological-mineralogical, structural, and geotectonic factors has been developed.

(5) It is shown that screening properties of clay caps at depths over 6.5 km will be preserved; and this fact together with good filtration-capacity parameters of reservoirs, abnormal high pore pressures, and relatively low temperatures, which make it possible to preserve hydrocarbons, allows a confident prediction of the probability of a discovery of commercial oil and gas accumulations at depths of 9 km and more in the Southern Caspian Depression.

REFERENCES

Belov, N.V., and others, 1978, Swelling mineral of deeply
    submerged Pliocene sediments of the Baku
    archipelago: DAN SSSR, v. 243, no. 3, p. 733-742.

Burst, J.F., 1969, Diagenesis of Gulf Coast clayey sediments and its
    possible relation to petroleum migration: Am. Assoc.
    Petroleum Geologists Bull., v. 53, no. 1, p. 73-93.

Buryakovsky, L.A., and Djevanshir, R.D., 1979, Characteristic of
    pore space structure of non-consolidated clays
    (productive series of the Baku Archipelago as an
    example): DAN Azerb. SSR, v. 35, no. 7, p. 70-74.

Buryakovsky, L.A., Djafarov, I.S., and Djevanshir, R.D., 1982,
    Prediction of physical properties of collectors of oil
    and gas caps: M., Nedra, 200 p.

Buryakovsky, L.A., Djevanshir, R.D., and Aliyarov, R.Yu., 1983,
    Problem of studying abnormally high geofluid
    pressures in connection with survey, exploration and
    development of oil and gas deposits: Izv. AN Azerb.
    SSR, ser. Nauki o Zemle, no. 1, p. 119-127.

Kheirov, M.B., 1979, Influence of depth of sedimentary rocks
    bedding on transformation of clayey minerals: Izv. AN
    SSSR, ser. geol., no. 8, p. 144-151.

Khitarov, N.I., and Pugin, V.A., 1966, Montmorillonite in
    conditions of increased temperatures and pressures:
    Geokhimiya, no. 7, p. 790-795.

Klubova, T.T., 1973, Clayey minerals and their role in genesis,
    migration and petroleum accumulation: M., Nedra,
    256 p.

Millot, J., 1968, Geology of clays (trans. from French): L., Nedra,
    360 p.

Narimanov, A.A., 1984, Oil and gas content perspectives of deeply
    bedded horizons in the western shelf of the southern
    Caspian Sea. Review information: M., VNIIEGasprom,
    is. 1, 38 p.

Powers, M.C., 1967, Fluid-release mechanisms in compacting
        marine mudrocks and their importance in oil
        exploration: Am. Assoc. Petroleum Geologists Bull.,
        v. 51, no. 7, p. 1240-1254.

Zkhus, I.D., and Bakhtin, V.V., 1979, Lithogenetic transformations
        in zones of abnormally high reservoir pressures: M.,
        Nauka, 150 p.

Zlochevskaya, R.I., 1976, Connate water in clayey soils: M., izd-vo
        MGU, 176 p.

GEOL: AN INTERACTIVE SYSTEM FOR DATA PROCESSING [1]

J.J. Royer [2], P. Jacquemin [2], and J.L. Mallet [3]

Centre National de la Recherche Scientifique, Nancy [2]
Ecole Nationale Supérièure de Géologie, Nancy [3]

ABSTRACT

Specific software for the processing of geological data is needed
to optimize geostatistical, automatic mapping, and data analysis.
The GEOL package is an interactive user-friendly system which is
concerned with structures and algorithms specially designed and
implemented to the handling of data encountered in geosciences.
Several commands can manipulate, visualize, and analyze the data
quickly from a database. The different recognized types of data
are numerical, alphanumerical, logical, or of set type. They can be
dimensionless or implicitly one-, two-, or three-dimensional
gridded.

The GEOL package is an open system comprising: an editor (GMS)
to execute calculus or any retrievals on variables in the GEOL Data
Table, an interpolation program (INT) to convert an irregular
observed data into a one-, or two-, three-dimensional grid, a
multivariable statistical analysis program (GMST) and an automatic
map-making program (MAP) to draw spatially distributed data as
contour maps, block diagrams or 3D contour surfaces.

The integration and user-friendliness of the system make it easy
to use even by beginners at computer science and adequate for
teaching data-processing concepts.

---

[1]CRPG Contribution No. 702

INTRODUCTION

Anyone whose experience of data processing in Earth sciences
has been limited to computer library ready-to-use packages could
be forgiven believing that the job of a computer geoscientist is
limited to translating data files produced by a given package to an
another file formatted for another user package. Generally
speaking, to perform a multivariable data analysis followed by
automatic mapping then by kriging estimation using published
software, much time will be spent to fit data output to data input.
This is the reason why a decision to design and develop an
integrated system was made by the CRPG in the 1980's.

The Project first started with the definition of concepts and
structure used in Earth-science data (see for instance Cunin and
others, 1979).

Then a specialized language was designed to simplify the
manipulation of structures which are complex from the point of
view of computer science but, nevertheless easy to use for
geoscientists (isovalue maps, grids, faults, set of data,....). the
essential characteristics and the basic ideas of this new language
named GEOL, have been published. The GEOL compiler as well as
self-taught packages are implemented on various time-share host
or macro virtual 32-bit computers. In a recent paper, the authors
analyzed and discussed the needs of developing modular user-
friendly software using the original concepts defined in the GEOL
compiler project. Because the software is interactive, special
structures and algorithms must be defined and implemented.

In this paper, the philosophy used to build interactive commands
is discussed.

BASIC DATA STRUCTURE OF THE GEOL SYSTEM

The GEOL system has been designed with a specific data
architecture: the GEOL Data Table, abbreviated to GDT, has been
optimized for large data access, retrieval, manipulation, and
calculations. These GDT are natural concepts that allow an easy
management of data characterized by geographical coordinates,
for instance involving various geomathematical fields such as
geoscience data analysis, mining studies, geography,
environmental survey, automated mapping, meteorologists,etc.

GEOL GDT Structures

The GDTs are organized in:

- GEOL components (for example columns or variables) consisting in a collection of information measured for a given parameter defined onto a geometrical structure (a claim, a district, a log, or a 3D volume). The parameters might be a single variable, for instance a grade or other measurement, a vectorial component, a characteristic function, or a set.

- statistical units (for example, rows) referring to sample points containing all the information available for a particular point of the studied area or spatial domain, that is the values of the whole components at this point.

In order to simplify the manipulation of complex information structures (sets, 3D grids, vectorial components, etc.) and to increase confidence in program execution, each object is characterized by a name identifier up to 8 characters and the type of structure.

Different Types of GDTs

GDTs differ according to the nature of the data. Four types are in use:

(a) Data Table type (DT) is dedicated to data measured on irregular or randomly spaced points.

(b) Data Table n type (DT1, DT2, DT3) are used when the data are recorded on the nodes of a regular grid in the 1,2, or 3D space. In practice, regular grids are estimated from irregular sampling points (DT) using an interpolated method.

In terms of computer objects, evaluating a DT2 or DT3 from a DT corresponds to a type conversion similar to those involved in converting a Real into an Integer. But, the process involved for the estimation of the grid is more complicated than a truncation. A specific command named INT (for INTerpolation has been designed to estimate a 1,2, or 3D grid from randomly spaced data points.

The advantages of using regular grids are:

    (i) the use of implicit coordinates;

    (ii) more efficient computation algorithms and graphical displays for interactive exploitation:

    (iii) a better data management.

The GDT type and the gridding description are stored when creating a file. Then, future uses are submitable to some checking for maximum safety.

Conversion of Type

The conversion of type consists in generating a GDT of type DTn (n=1 to 3) from a GDT of type DT by an appropriate numerical approximation. The values of a given parameter are computed at each node of a profile (DT1), a surface (DT2), or function (DT3) defined respectively onto a linear, rectangular, or parallelepipedic grid. Several algorithms defined by options are available: trend surface, splines, least squares, and kriging. The numerical approximation can be done taking into account discontinuities (faults, borders, etc.) in a given area defined by a set of polygonal contours in 2D or a set of triangular facets in 3D. Methods for the estimation of a boolean, alphanumerical, or real variable have been studied specially in order to provide fast interactive algorithms.

THE GEOL LANGUAGE

A specialized language has been developed for manipulating and maintaining the structures of the database. It offers easy mapping and computing facilities for a user to develop software using the GDT facilities. It provides: (i) implicit use of GDT structure treated as data libraries with automatic transfer between primary and secondary or virtual memory; (ii) structured programming with loops, tests, and subroutines; and (iii) grid to grid applications.

Procedures are available which present results as graphs, contour maps, or in other pictorial forms. The GEOL compiler implemented in a Unix environment has been written in FORTRAN for portability reasons. The programs generated by the compiler are FORTRAN code and can be used themselves as packages or incorporated in other general purposes programs. Other possibilities are offered: (i) insertion of FORTRAN statements in a GEOL program; (ii) use of GEOL subroutines from the user's library, or the GEOL system library, but also of FORTRAN subroutines from an existing library.

The GEOL language is similar to a FORTRAN preprocessor such as RATFOR. But, the main differences are: (i) this is a structured language with pretyped remnant objects: grids, set, list of variables, vectorial components (variables), and (ii) specific operators (union, intersection of sets). A detailed description of the GEOL language is contained in Bouchet and others, 1980, Bouchet, 1982, and Royer and Mallet, 1980.

GEOL CONVERSATIONAL PROGRAMS.

A conversational program must be self-documented and self-taught providing help for the new or unfamiliar user. It is no use to read hundreds of pages before obtaining a result. It also must be fast and easy to use; it must be robust to mistakes. No crash if the user answers incorrectly, but it must control and test inputs in order to increase the safety of exploitation. It also must be fair and polite.

In order to satisfy these conditions, all the GEOL conversational programs have been designed in a similar framework. For instance, at each step the user has the following possibilities:

   (i) to work the program backward by entering a carriage return (CR), for instance to change a previous input or option, or forward by entering a proper input;

   (ii) to ask for explanations by typing a question mark in order to edit a manual;

   (iii) to stop the program by the reserved key word STOP, the effect of this general command is to stop the program in a safe way, that is saving GDT in the database after validation by the user.

Playback files have been used extensively in a way similar to the use by the Unix system. Each input and output are recorded in a specific file. These files can be run by the user for similar or repetitive tasks. The advantages are: (i) to keep a trace of the work; (ii) to optimize tasks whose inputs are similar - for instance if one wants to create a grid on several components (variables) using a similar spline method, the neighborhood, the degree of the spline, the geometrical set on which the interpolation is performed, are defined only once, then saved in a playback file and when reused for the other components. Using the other facilities of the Unix system, the playback files facilities provide the possibility of creating preparameter commands; and (iii) to constitute a type of library standard inputs (predefined options).

Other techniques can be used to speed the dialog between the user and the program. For instance, selection of options by key words is used extensively rather than questions such as "Do you want to perform such things, yes or no?". Syntactic structures may be used to get the option the user needs. For instance the sentence: SMTH GRD2 DBL using the contouring programs indicates that the user wants to draw a double contour line (DBL) after smoothing the grid values (SMTH) and writing the isovalues using two decimal along the isolines.

New users would certainly use less sophisticated answers by asking help (?) to enter the appropriate options.

These programs need virtual memory because all the data are copied into the main memory before proceeding.

The advantages are: fast access in the database because the computer optimizes the input or output using virtual memory algorithms; safety of the data, the resulting GDT is recopied onto the old version only at the end of the session; rapidity of execution.

The disadvantages are: space needs for large GDT; it takes time for reading and saving the GDT.

Moreover, numerical algorithms have been studied extensively in order to optimize the computation time and the user can obtain a 3D representation in few minutes. But the quantity of programming work involved in the optimization of dialogs is equivalent to that necessary for the numerical processing.

The conversational programs of the GEOL system are dedicated to special tasks such as: easy use of the GEOL compiler (program),

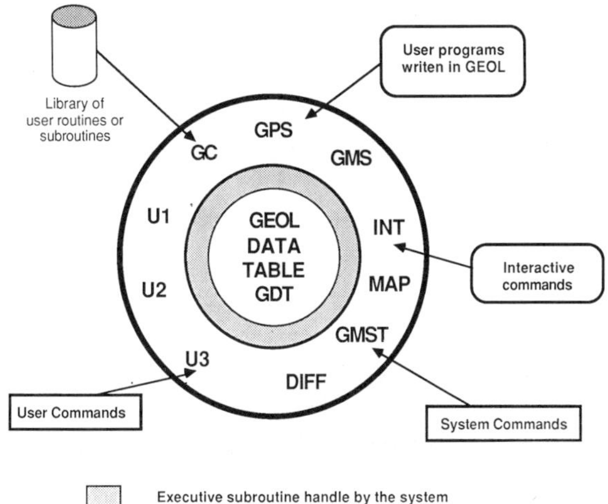

Figure 1. Structure of GEOL system.

interactive management system (GMS program), gridding,
automated mapping, and other applications.   These programs
offer implicit access to the data stored in GDTs. Figure 1
summarizes the structure of the GEOL system and the
conversational programs.

The GEOL compiler (GC): is a software written in the C language,
FORTRAN, and UNIX. Its aim is to help the user for running their
own GEOL programs. GC facilities are: compilation and execution
of programs written in GEOL language; creation and management
of GEOL libraries and subroutines; and creation and management
of data files and GDTs called by the input program.

The GEOL management system (GMS): is a conversational
management and retrieval program for user's GDTs including
creation, modification, extension, or destruction of a GDT as well
as components. Editing facilities allows the modification of the
data stored in a GDT component. This edition should be done
line by line, or using a conditional set or the coordinates of the
grid nodes in a GDT situation. An interpreter including most of
the facilities provided by the GC also is available. It can be used to
created new components defined by a mathematical formula or by
a condition (for set components). The calculation could be
restricted onto a given set or for all nodes of the grid.

INT program (the interpolation program) estimates the values of an irregular nongridded (x,y,z) function onto a gridded form defined in 1,2, or 3D. Several numerical methods are proposed, such as least square, splines, single kriging, or BLUEPACK methods. The estimation methods take into account the discontinuities of the studied function (for example faults, discordances in geology, limits of claims, etc.). The estimation variance or jacknife variance also can be evaluated. The input data and the discontinuities are stored in DT files; the output estimated values are saved in a DT1, DT2, or DT3 file according to the nature of the studied function. The estimation can be restricted to a geometrical set (claims, borders, etc.).

MAP program is an automated contouring conversational program including faults of discontinuities handling in 2 or 3D and usual representation of gridded or plot function onto a geometrical set or inside a polygonal contour. Large number of possible color displays for one of more functions are available: 2D contour line maps, block diagrams, zoning maps, cross sections, or 3D isolevel in perspective, 3D cross sections along planes, etc.

Examples of output using the MAP program are shown in Figures 2 to 6.

GMST is a conversational program built around a unique algorithm covering the different methods used in factor analysis including R-mode, Q-mode, discriminant, or generalized correspondence analysis. Different display facilities are provided to represent cross-plots, histograms in 1D or 2D, distribution functions, or factor analysis cross-plots.

Other commands including (i) resolution of the diffusion equation by finite-element algorithms using direct search methods and inverse problems (DIFF); (ii) factorial spatial analysis such as proximity analysis; (iii) linear quadratic problems subjected to linear constraints (QNORM) also are available. Applications in geothermics, fluid migration through porous media, geochemical analysis, or norm calculation also can be handled.

Other commands make the interface between the GEOL database and commercial packages extensively used in the mining or petroleum industries, that is BLUEPACK, SIMPACK, MAGMA, or volumetrics programs.

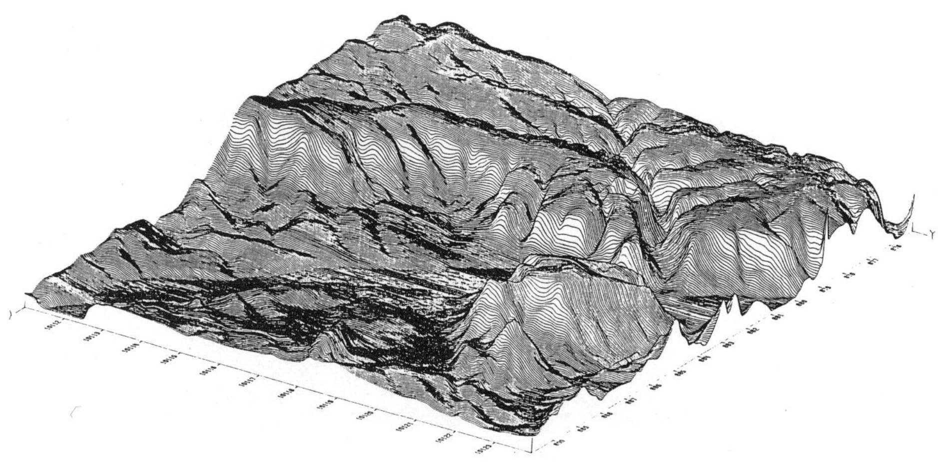

Figure 2.   Topographic map of Nancy District, (France);
vertical scale is exaggerated.

CASE STUDY

Different examples taken in geoscience, are given in Figures 2 to
7 in order to illustrate the possibilities of the MAP program.

DISCUSSION

The GEOL system does not provide a general database
management system, rather it provides a simple structure for
processing data.

This system is an open one in the sense that its compiler provides
user programming tool for their own development. For instance,
an interface with the BRGM database has been developed.

The GEOL system is easy to use after some training even by
beginners in computer science and is adequate for teaching data-
processing concepts. The self-taught facilities mentioned here
reduce the training costs.

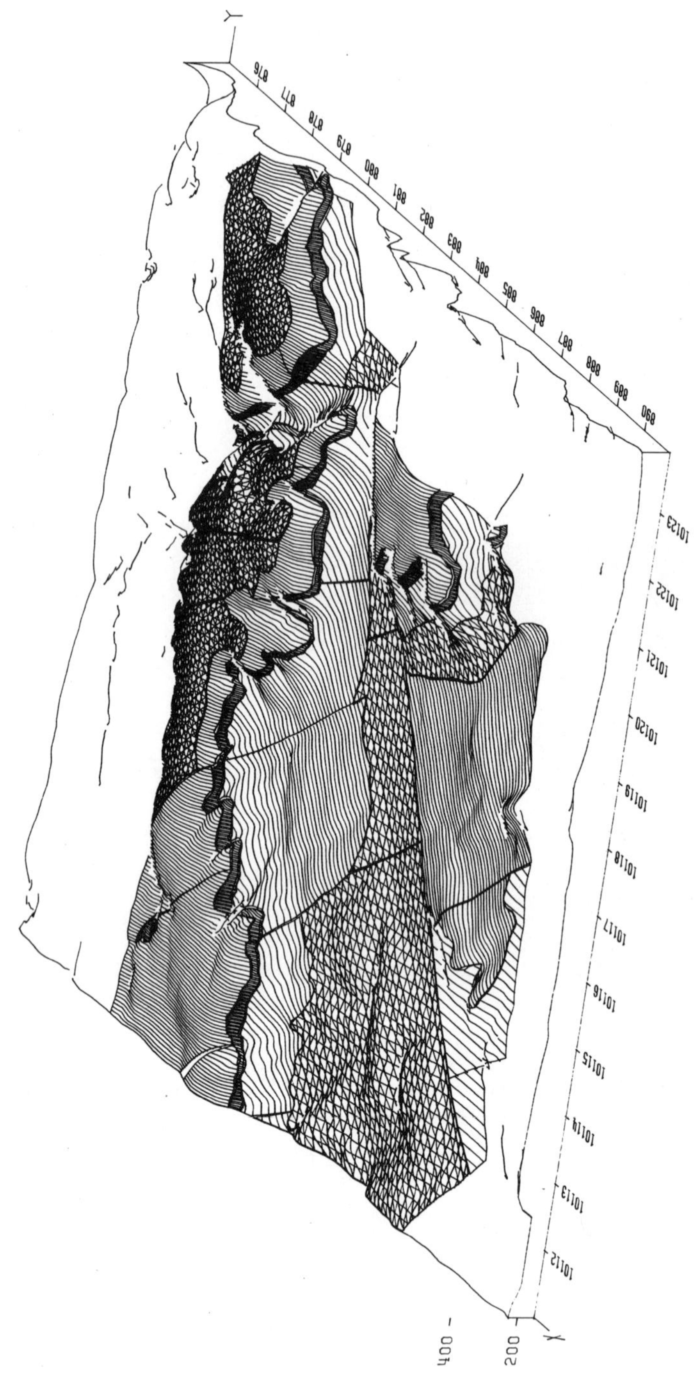

Figure 3. Same topographic block diagram of Nancy District on which geological map has been projected.

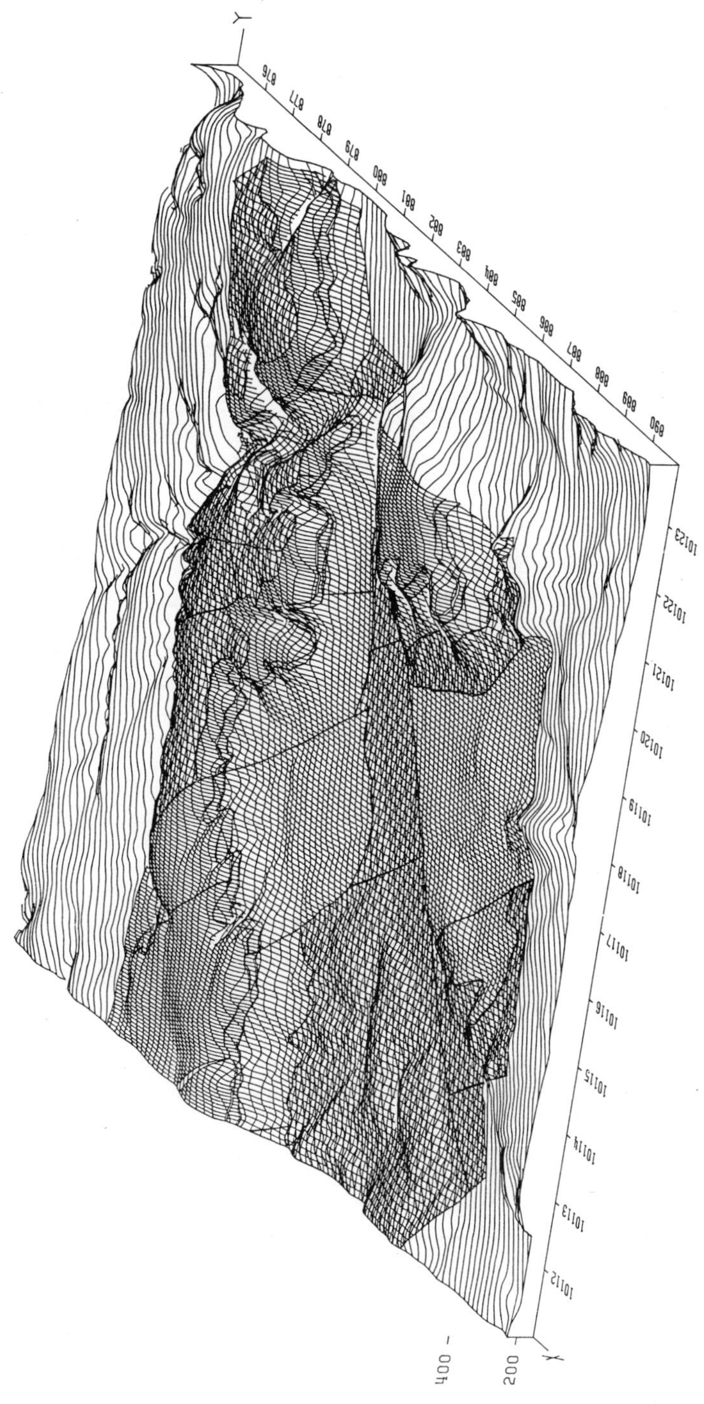

Figure 4. Geologic map drawn onto topographic block diagram.

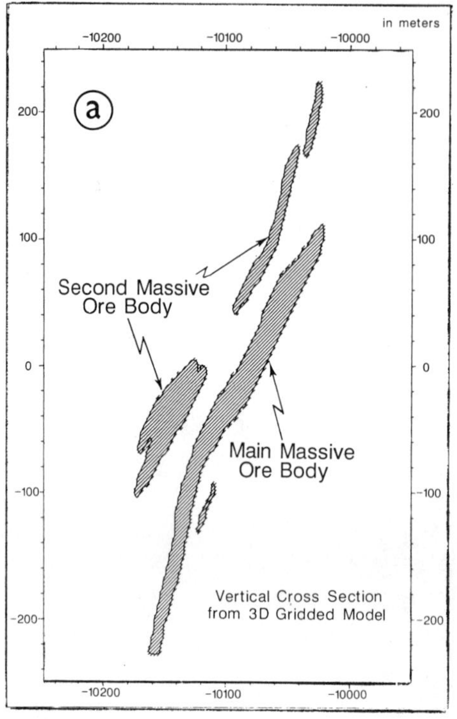

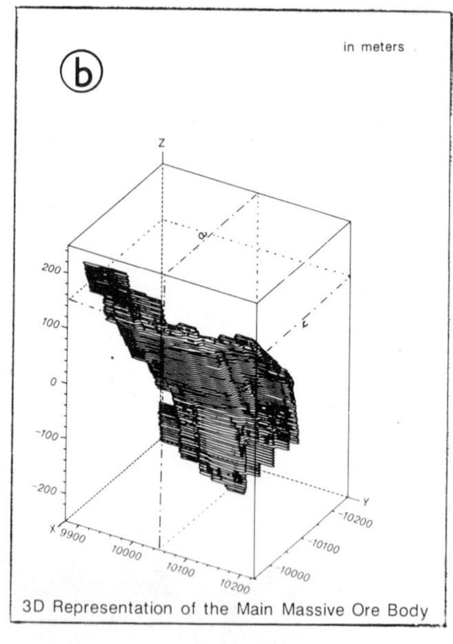

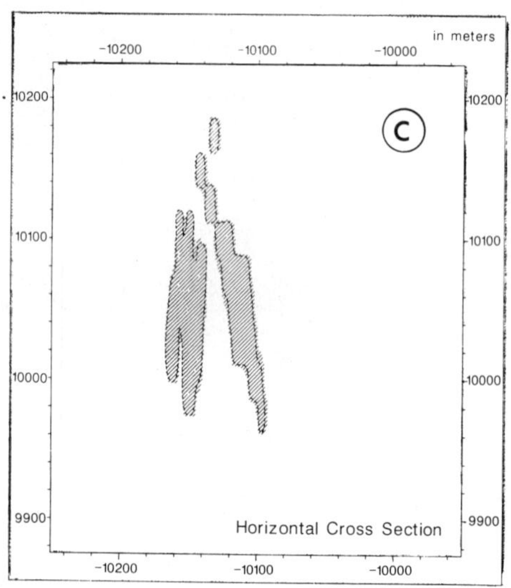

Figure 5. 3D representation of volcano-sedimentary deposit: (a) vertical cross-section obtained from 3D-gridded model; (b) principal orebody represented by horizontal slices; and (c) horizontal cross-section.

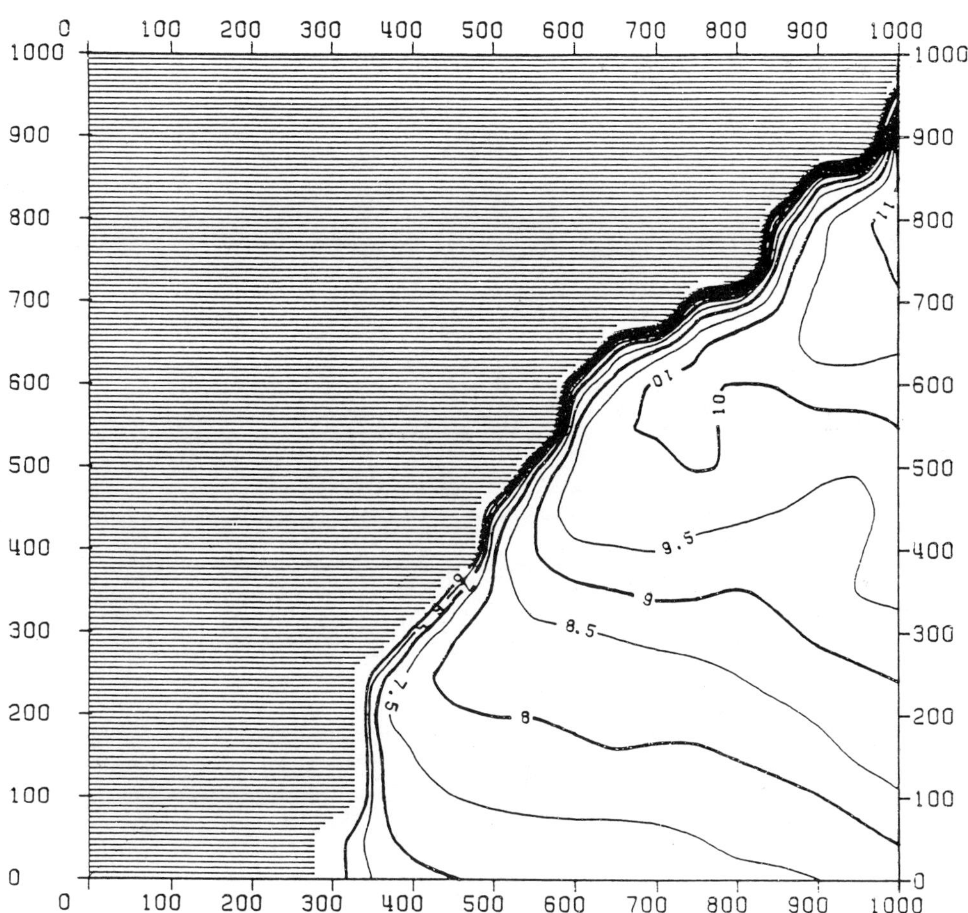

Figure 6a. Amount in ore of seam; contour line map.
Shaded zone shows part of studied domain where
seam outcrops.

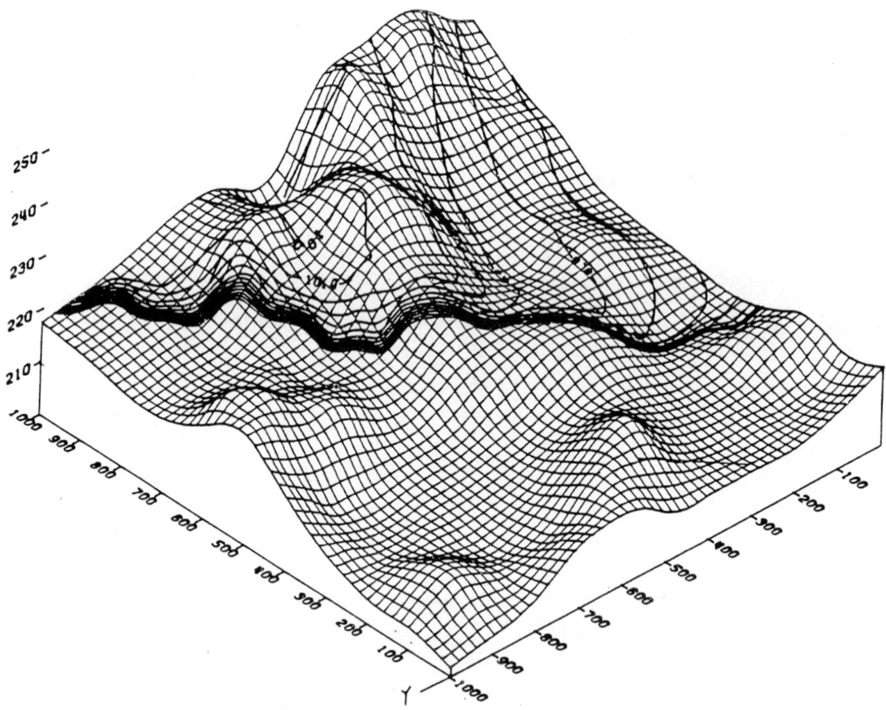

Figure 6b. Contour line map of amount in ore displayed on perspective view of topography.

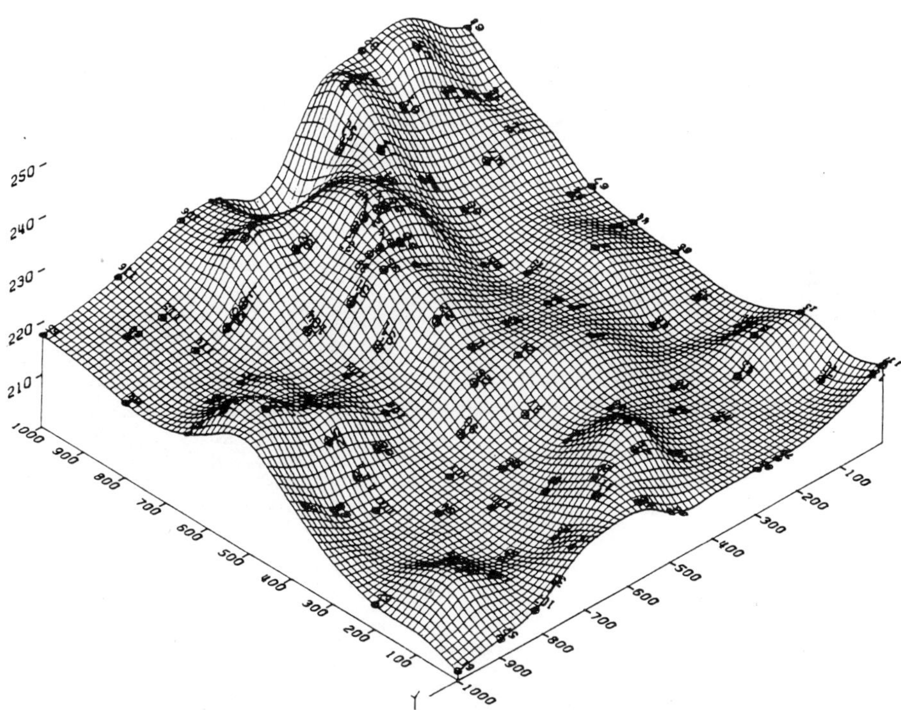

Figure 7a. Studied domain. Perspective view of topography of explored zone showing location of sampling points.

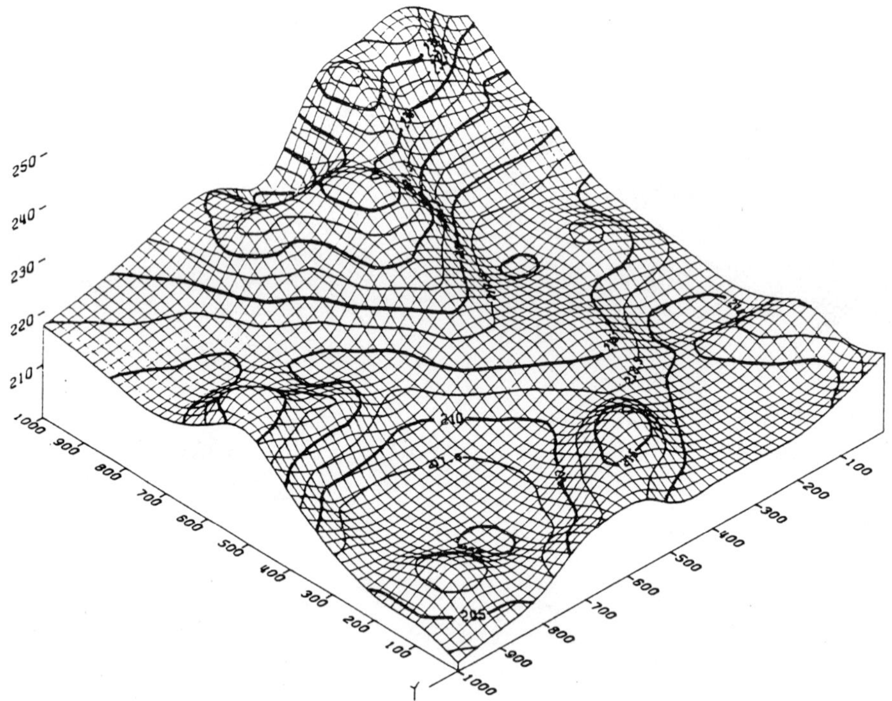

Figure 7b. Studied domain. Perspective view of topography
of explored zone. Mesh perspective and contour
lines are displayed from estimated grid.

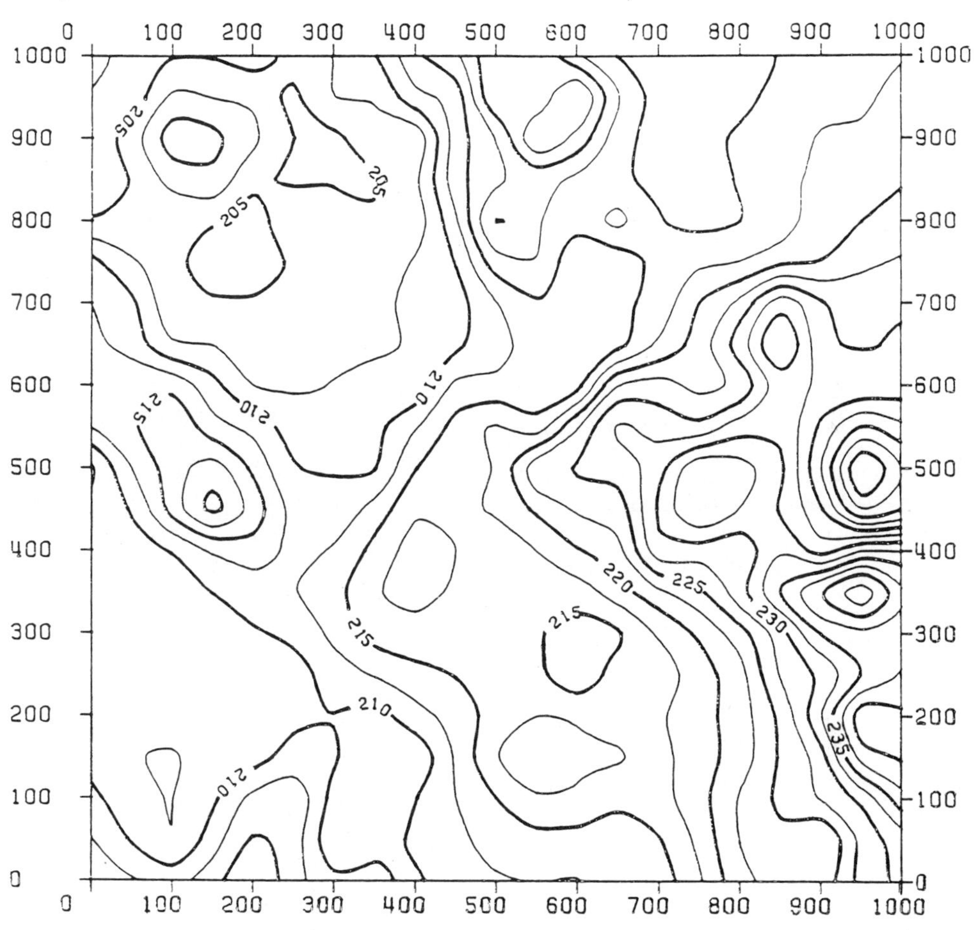

Figure 7c. Contour-line map of topography drawn from
50 x 50 m square grid.

ACKNOWLEDGMENTS

CISI Petrol Services Company is acknowledged for providing some of the case studies.

REFERENCES

Bouchet, P., Cunin, P.Y., Griffiths, M., Mallet, J.L., and Royer, J.J., 1980, GEOL: a graphical and statistical language for geodata management: Compstat, Physica-Verlag, Vienna for IASC, p. 88-94.

Bouchet, P., 1982, Definition et implantation d'un language de manipulation de données: unpubl. masters thesis, INPL, Nancy (France), 121 p.

Cunin, P.Y., Griffiths, M., Mallet, J.L., and Royer, J.J., 1979, GEOL: langage d'analyse de données géologiques: Congrĕs de la Convencion Informatica Lătina, Barcelone, 5 p.

Jacquemin, P., Mallet, J.L., and Royer, J.J., 1985, Interactive computer aid design in the processing of mining and geological data, in Glaeser, P.S., ed., The role of data in scientific progress: Elsevier Science Publ. B.V. (North-Holland), Amsterdam, p. 19-24.

Royer, J.J., and Mallet, J.L., 1980, Recent progress in spatial data processing: The GEOL language: Science da La Terre, Ser. Informatique Geologique no. 14, Nancy (France), p. 177-190.

NUMERICAL CLASSIFICATION OF MESOZOIC VOLCANIC ROCKS IN THE EASTERN PART OF CHINA AND ITS GEOLOGICAL SIGNIFICANCE

Liu Chengzuo and Chai Junjie

Academia Sinica

ABSTRACT

In this paper a numerical classification of Mesozoic volcanic rocks in eastern China has been approached by multivariate methods and its geologic significance is discussed. Because of complexity of geologic conditions, representative regions, and typical rock specimens are selected based on petrological and geomathematical research. For the purpose of numerical classification the factor analysis, cluster analysis, discriminant analysis, and nonlinear mapping are utilized. The obtained results verify the rationality of division of these volcanic rocks into groups from their rock-type characteristics, and present much useful information which may be utilized for theoretical and practical purposes.

GEOLOGICAL CONDITIONS AND PRESENTATION OF THE PROBLEM

Mesozoic volcanic rocks are distributed widely in seventeen provinces in the eastern part of China and constitute an important component of volcanic belt surrounding the Pacific Ocean. The prevailing part of them consists of volcanic rocks of continental facies. Because of their close relationship with various metal and nonmetal deposits, many theoretical and practical problems concerning them have attracted attention of the Chinese geologists.

Mesozoic volcanic rocks have been approached by the Chinese geologists in various ways. Among them, Prof. Wu Liren has studied Mesozoic volcanic rocks in the eastern part of China from petrological point of view and has divided volcanic rocks into the southern, central, and northern petrological regions in accordance with their petrological association. The main boundaries between these regions are the Changzhou-Yueyang Fault in the south and Baodi-Changli-Jinzhou Fault in the north, which extend in a roughly east-northeast direction.

In the southern region andesite-(dacite)-rhyolite is the main volcanic rock association, with rhyolite as the predominant member, in which alkali-feldspar rhyolite occupies an important position. But basalts (tholeiite, trachy-basalt) occur locally. Besides basic volcanic rocks and andesites that belong to the alkali rock series, all others are referred to the calc-alkali rock series.

In the central region potassium-rich volcanic rock associations, such as (trachybasalt)-trachandesite-latite-trachyte-(phonolite), are well developed. They are referred to alkali rock series, straddle -A type (Ningwu, Luzong Basins) or -B type (East Shandong, Central Shandong) of differentiation trend. They occur in or along the rift zones.

In the northern region, the main volcanic rock association is composed of basalt-andesite-dacite-rhyolite and belongs to calc--alkali rock series, but quartz trachybasalt-quartz trachandesite-quartz latite-rhyolite also may occur, especially in the Yanshen area. In this situation they are referred to a transitional type from alkali to calc-alkali rock series.

The composition of all the volcanic rocks is characterized by high alkalis, especially potassium, and the ratio $K_2O/Na_2O \geq 1$, but low titanium (TiO_2 usually $< 1\%$).

Research of Mesozoic volcanic rocks in the eastern part of China started at the end of the 1920s. Prior to that time only a few research works were done there. Later, geologic data were collected in great quantities and studied by experts. Therefore, important results have been obtained in recent years.

Nevertheless, many theoretical and practical problems of Mesozoic volcanic rocks being connected with mineralization are not settled yet and more detailed and thorough research is necessary.

Furthermore, one point should be stressed: the previous results are of a qualitative character and were accomplished by traditional methods. In the past few years, development of mathematical geology provides evidence that quantitative methods have certain advantages over qualitative ones. Therefore, problems of approaches of Mesozoic volcanic rocks by geomathematical methods have been presented. The senior author of this paper has dealt with these problems and some results have been obtained (Liu, 1979a, 1979b; Liu and Chai, 1982). On the basis of previous works, some new progress of numerical classification of Mesozoic volcanic rocks is made and its geologic significance is discussed.

CONCEPT AND MATHEMATICAL MODELS OF NUMERICAL CLASSIFICATION

Concept of Numerical Classification

For adaptation of special features of the problem, the idea of numerical classification of Mesozoic volcanic rocks includes the following aspects:

> (1) selection of representative research subjects; and

> (2) classification of representative research subjects numerically by a synthetic utilization of several multivariate methods.

Because of the complexity of geologic conditions, it is necessary to pick out representative regions and typical rock specimens, based on combination of petrological and geomathematical methods. For this purpose, a multivariate procedure for evaluation and improvement of classification is applied (Demirmen, 1969). Results of petrological research, obtained by Prof. Wu Liren (Wu and others, 1982a, 1982b, 1984) are accepted as the geologic foundation for selecting the research subjects.

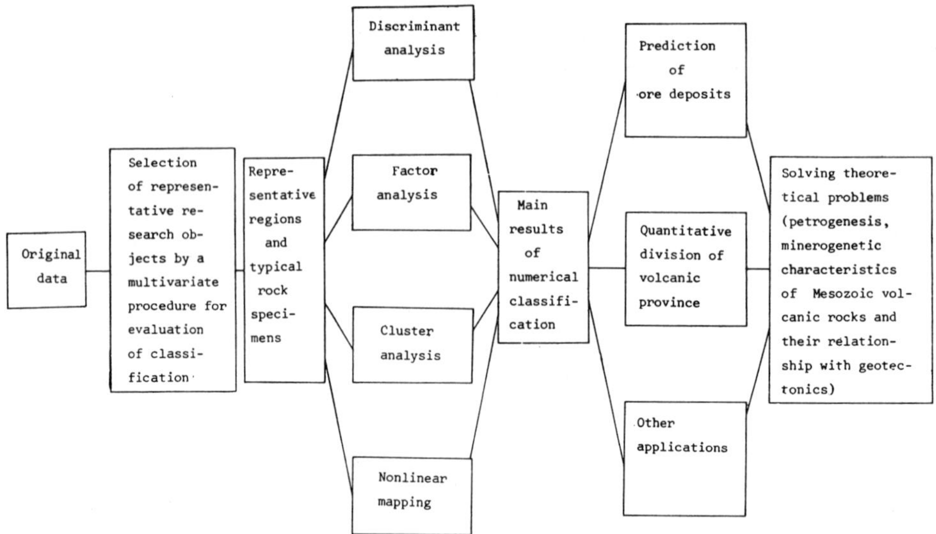

Figure 1: Scheme of sequential course of research works.

Experience of geomathematical research shows that every multivariate method possesses its own special merits and demerits. Thereby, a synthetic utilization of several multivariate methods may compensate their disadvantage and provide satisfactory results. In this paper a numerical classification is realized by discriminant analysis, factor analysis, cluster analysis, and nonlinear mapping. The obtained results are used for some theoretical and practical purposes. Figure 1 shows schematically the sequential course of research that joins qualitative and mathematical models of numerical classification.

Mathematical Models of Numerical Classification

In our research multivariate procedures for evaluation and improvement of classification (Demirmen, 1969; Davis, 1973; Liu and Sun, 1981; Wilks, 1932) are used for selecting typical specimens of representative regions. Process of such a selection contains the following steps:

(1) Partition of rock specimens into preliminary groups according to the three petrological regions (Wu, 1982). In this paper we divide our rock specimens into three preliminary groups: the first group - Fujian, Zhejiang, and Guangdong Provinces (22 andesite specimens); the second group - the northern part of Hebei

Province (31 andesite specimens) and the third group - Jiangsu and Anhui Provinces (39 andesite specimens). A total 92 andesite specimens were investigated.

(2) Evaluation and improvement of preliminary groups by multivariate procedures. Evaluation of preliminary groups is made on the basis of three criteria (trace W criterion, Wilks' lambda criterion, and trace $W^{-1}B$ criterion). The three criteria measure the degree of "compactness" of a classification and the meaning of "compactness" will be evident in a discussion of the criteria.

The three criteria of evaluation as defined are based on the within, between, and total scatter matrices. Let

$$X_{h.i} = \frac{1}{n_h} \sum_{k=1}^{n_h} X_{hki}$$

be the mean of the i-th variable (X_i) over the h-th group, and

$$X_{..i} = \frac{1}{n} \sum_{h=1}^{m} n_h$$

$$X_{h.i} = \frac{1}{n} \sum_{h=1}^{m} \sum_{k=1}^{n_h} X_{hki}$$

be the grand mean of the i-th variable of the n item. Then the matrices

$$\underline{W} = (w_{ij}), \quad \underline{B} = (b_{ij}), \quad \underline{T} = (t_{ij}), \quad i, j = 1, \ldots, p. \text{ where}$$

$$w_{ij} = \sum_{h=1}^{m} \sum_{k=1}^{n_h} (X_{hki} - X_{hi})(X_{xkj} - X_{h.j}),$$

$$b_{ij} = \sum_{h=1}^{m} n_h (X_{h.i} - X_{..i})(X_{h.j} - X_{.j}), \text{ and}$$

$$t_{ij} = \sum_{h=1}^{m} \sum_{k=1}^{n_h} (X_{hki} - X_{..i}) (X_{hki} - X_{..j})$$

represent, respectively, the within, between, and total scatter matrices in the X^- space. Furthermore, $\underline{T} = \underline{W} + \underline{B}$.

Trace W criterion

$$\text{tr } \underline{W} = \sum_{h=1}^{m} \sum_{k=1}^{n_h} \sum_{i=1}^{p} (X_{hki} - X_{h.i})^2$$

is the within-group sum of squares with respect to all p variables pooled over all m groups.

Wilks' lambda criterion is

$$\Lambda = \frac{|\underline{W}|}{|\underline{T}|}$$

Λ represents the ratio of within-group scatter to total scatter.

Trace $W^{-1} B$ criterion is as follows

$$\text{tr } \underline{W}^{-1} \underline{B} = \sum_{h=1}^{m} n_h \sum_{i,j=1}^{p} w^{ij} (X_{h.i} - X_{..i}) (X_{h.j} - X_{..j}) ,$$

where w^{ij} is (i,j)-th element of $\underline{W}^{-1}$. A classification with smaller values of tr $\underline{W}$ or Λ and larger values of tr $\underline{W}^{-1}\underline{B}$ indicate a more "compact" or "better" classification.

The improvement of a classification is performed in the discriminant space. Furthermore, an efficient method of improving a classification by the criteria is provided by the nearest-neighbor algorithm. The utilized program allows the algorithm to operate in the discriminant space generated from the initial space. Iterations are terminated when an improvement by the nearest-neighbor algorithm is no longer possible, or when the maximum number of iterations specified by the user is exceeded.

(3) Selection of typical rock specimens of representative regions. After application of the described procedures, we regard the rock specimens, which are taken from the northern, central, and southern petrological regions and have been identified as specimens of their own regions, as typical specimens of representative regions. The remaining rock specimens are used for other purpose, which will be explained in the next section.

Beside the described method, other conventional multivariate methods such as: discriminant analysis, factor analysis, cluster analysis, and nonlinear mapping are used as mathematical models for numerical classification. In this paper we use these models in their currently utilized forms.

MAIN RESULTS OF NUMERICAL CLASSIFICATION

Result of Selection of Representative Regions and Typical Rock Specimens

Using methods explained in the previous section the following 64 andesite rock specimens from Fujian and Guangdong Provinces; 26 andesite specimens from the northern part of Hebei Province; and 22 andesite specimens from Jiangsu and Anhui Provinces as typical rock specimens of the southern, northern, and central petrological regions, respectively.

Regarding andesite specimens as typical ones is due to their importance in magmatism, close relationship with iron deposits and their widespread distribution in the seventeen provinces in the eastern part of China. Percentage content of ten oxides are selected as argument in our research. The arguments of typical specimens in representative regions are shown in Table 1.

In further investigations, we regard the typical rock specimens as representative of Mesozoic volcanic rocks in the eastern part of China and they serve as object of multisided and thorough research.

Table 1. Mean of argument of typical specimens in representative regions

Mean Percentage Content of Oxides

Representative Regions	SiO_2	TiO_2	Al_2O_3	Fe_2O_3	FeO	MnO	MgO	CaO	Na_2O	K_2O
Fujian and Guangdong prov.	59.22	1.07	16.25	2.50	4.98	0.11	2.94	5.93	3.22	2.22
Northern part of Hebei prov.	55.79	1.43	15.69	6.96	1.99	0.27	1.80	4.09	4.49	3.10
Jiangsu and Anhui prov.	56.72	0.77	16.71	4.73	2.71	0.15	2.18	4.27	3.94	3.80

Result of Discriminant Analysis

Discriminant functions are established for 64 typical andesite specimens by multiple stepwise discriminant analysis. Coefficient of discriminant functions for three petrological regions are shown in Table 2.

The discriminant functions consisting of the coefficient from Table 2 are only the preliminary and they may be improved with collection of more complete original data and further on-going investigations.

Results Obtained by Factor Analysis, Cluster Analysis, and Nonlinear Mapping

For convenience of research works the 64 typical andesite specimens are merged into 30 groups, which consist of andesite specimens with similar content and nearby occurrence. According to results obtained by factor analysis, cluster analysis, and nonlinear mapping, 30 groups of andesite specimens are divided into four combinations: three combinations corresponding with the three petrological regions and the fourth combination possesses miscellaneous character. Besides andesites, specimens

Table 2. Coefficient of discriminant functions for three petrological regions.

Discriminant Coefficients of Arguments

Petrological Regions	Fe_2O_3	K_2O	TiO_2	SiO_2	MgO	Na_2O	CaO	FeO	MnO	Al_2O_3	Zero term
Fujian & Guangdong prov.	27.06	103.7	26.00	27.43	11.58	45.63	11.31	38.95	31.14	28.92	-1347.91
Northern part of Hebeiprov.	26.84	113.55	24.07	31.22	9.20	40.58	6.34	42.00	38.97	35.15	-1376.01
Jiangsu & Anhui prov.	26.10	101.56	25.02	29.65	10.66	40.50	8.25	40.16	34.92	33.07	-1302.23

of Mesozoic rhyolite, basalt, latite, trachy-andesite, and trachyte are studied using the same methods.

Summary of Results Obtained by Multivariate Methods

(1) Using multivariate methods we have selected typical specimens of representative regions and numerically classified them into three groups (belonging to the three petrological regions by Wu Liren), each of which has its own character connected with a degree of richness in alkaline and mineralization. Mean content of Na_2+K_2O is 5.44%, 7.59%, and 7.74% for the typical andesite specimens of southern, northern, and central regions, respectively.

(2) Results of geologic and geomathematical research show that the andesites of the southern region are not related to iron mineralization and that of the central region are connected closely with iron mineralization and the northern region possesses a transitional character.

(3) Besides, the obtained results present other useful information which may be utilized for quantitative division of volcanic provinces and other theoretical and practical purposes.

APPLICATIONS OF RESULTS OF NUMERICAL CLASSIFICATION

Prediction of Iron Deposits

Mineralization of iron deposits is related closely with geologic structure, lithological character of magmatic rocks, adjoining rocks, and many other factors. Furthermore, lithological character is an important factor influencing iron mineralization obviously. Results of numerical classification show close relationship of Mesozoic volcanic rocks of the central region with iron mineralization and it may be utilized for iron-deposit prediction. In our research, typical andesite specimens are considered as controlled samples and other specimens as unknown samples. By multiple discriminant analysis, major unknown andesite specimens occurring in the southern and northern regions may be identified as rock specimens of their own regions and the remaining unknown rock may be identified as specimens "occurring" in the central region which is related closely with iron mineralization. Therefore, the latter offers information for ore deposit prediction. For verification of rationality of such an inference, 8 unknown samples are studied by this method and the result shows a state of the affirmative circumstances. Similar results may be used for making ore prediction maps on small scales.

Quantitative Division of Volcanic Zones

Using the information extracted from results of multivariate analyses, areas where Mesozoic volcanic rocks are well developed may be divided quantitatively into volcanic zones. The main bases of quantitative division of volcanic zones are:

(1) Results of geomathematical research including all useful information presented by multivariate analyses;

(2) Division of volcanic zones by traditional methods from geologic and petrologic points of view; and

(3) Geologic conditions of interested regions.

Quantitative division of volcanic zones gives good results. The southern, northern, and central regions are divided into 7, 2, and 5 volcanic zones respectively.

CONCLUSIONS

In conclusion the following points should be stressed. First,
because of Mesozoic volcanic rocks of the eastern part of China is
a vast field for geologic investigations, the authors have done only
some preliminary work in one of its aspects. Second, despite all
this, some encouraging results are obtained such as selection of
typical specimens of representative regions, numerical
classification of them and other related rock samples, applications
of results for solving theoretical and practical problems. Third,
for further and more complete investigations of the problem,
some on-going work should be considered, such as an increase of
quantity of rock specimens and data, strengthening of data,
revising discriminant functions based on new data, and utilizing
obtained results for further theoretical and practical
investigations.

REFERENCES

Davis, J.C., 1973, Statistics and data analysis in geology: John
        Wiley & Sons, New York, 550 p.

Demirmen, F., 1969, Multivariate procedures and FORTRAN IV
        program for evaluation and improvement of
        classification: Kansas Geol. Survey Computer Contr. 31,
        50 p.

Liu Chengzuo, 1979a, Geomathematical researches of Mesozoic
        volcanic rocks in eastern part of China (in Chinese):
        Paper presented at Symposium on the sciences of iron
        deposits, Beijing.

Liu Chengzuo, 1979b, Numerical classification of Mesozoic
        volcanic rocks in eastern part of China and its
        relationship with iron deposits (in Chinese): paper
        presented at Symposium on sciences of iron deposits,
        Beijing.

Liu Chengzuo, and Sun Huiwen, 1981, Basic methods of
        mathematical geology and their applications (in
        Chinese): Publishing House of Geology, Beijing, 488 p.

Liu Chengzuo, and Chai Junjie, 1982, Quantitative research of
        Mesozoic andesites of the eastern part of China (in
        Chinese with English abstract): Scientia Geologia
        Sinica, No. 4, p. 415-420.

Wilks, S. S., 1932, Certain generalizations in the analysis of
        variance: Biometrika, v. 24, p. 471-494.

Wu Liren, and others, 1982a, Characteristics and genesis of the
        Mesozoic volcanic rocks in the eastern part of China
        (in Chinese with English abstract): Petrological
        Research (1), Geological Publishing House, Beijing,
        p. 11-27.

Wu Liren, and others, 1982b, Mesozoic volcanic rocks in the
        eastern part of China (in Chinese with English
        abstract): Acta Geologica Sinica, v. 56, no. 3,
        p. 223-234.

Wu Liren, and others, 1984, Mesozoic and Cenozoic rocks in East
        China and its adjacent areas (in Chinese): Science
        Press.

INDEX